Columba Sara Evelyn (Ed.)

Lavasa

Columba Sara Evelyn (Ed.)

Lavasa

Political corruption, Ministry of Environment and Forests, Hill station

Fec Publishing

Imprint

All parts of this book are extracted from Wikipedia, the free encyclopedia (www.wikipedia.org).

You can get detailed informations about the authors of this collection of articles at the end of this book. The editors (Ed.) of this book are no authors. They have not modified or extended the original texts.

Pictures published in this book can be under different licences than the GNU Free Documentation License. You can get detailed informations about the authors and licences of pictures at the end of this book.

The content of this book was generated collaboratively by volunteers. Please be advised that nothing found here has necessarily been reviewed by people with the expertise required to provide you with complete, accurate or reliable information. Some information in this book maybe misleading or wrong. The Publisher does not guarantee the validity of the information found here. If you need specific advice (f.e. in fields of medical, legal, financial, or risk management questions) please contact a professional who is licensed or knowledgeable in that area.

Cover image: www.ingimage.com
Concerning the licence of the cover image please contact ingimage.

Publisher:
Fec Publishing is a trademark of
International Book Market Service Ltd., 17 Rue Meldrum, Beau Bassin, 1713-01 Mauritius
Email: info@bookmarketservice.com
Website: www.bookmarketservice.com

Published in 2011

Printed in: U.S.A., U.K., Germany. This book was not produced in Mauritius.

ISBN: 978-613-7-46627-8

Contents

Articles

References

Lavasa

Lavasa	
— city —	

Lavasa

Location of Lavasa
in Maharashtra and India

Coordinates	18°24′19″N 73°30′23″E
Country	India
State	Maharashtra
District(s)	Pune
Time zone	IST (UTC+05:30)
Area	100 square kilometres (39 sq mi)
Website	www.lavasa.com [1]

Lavasa (Marathi: लवासा) is a private,[2] planned city being built near Pune.[3]

A 25000 acres (100 km^2)[2] or 8000 acres (32 km^2)[4] project being developed by HCC,[5] this as-yet-incomplete city has been controversial for multiple reasons including procurement of land,[6] harm to the environment,[2] and loans acquired through political corruption.[7]

In late 2010 the Indian Environment and Forests Ministry requested that construction cease because the project violated environmental laws.[8]

History

Billed as India's first hill station since Independence,[9] this private city is being constructed by a unit of Ajit Gulabchand's Hindustan Construction Company, Lavasa Corporation.[10]

Fortune Hotel, Lavasa

Land

The area demarcated for Lavasa is located near Pune in the Mose valley[6] of the Western Ghats.[11]

Lavasa Corporation

Currently majority owned by Hindustan Construction Company,[12] Lavasa Corporation is responsible for the construction of the city. It is preparing to make an initial public offering,[12] although this has been temporarily postponed.[13]

Legal name

Originally registered as Pearly Blue Lake Resorts Pvt Ltd in 2000, the company changed its name to The Lake City Corporation Pvt Ltd later that year and around 2004 changed it again to its current name, Lavasa Corporation.[14]

Ownership

As of late 2010 ownership of Lavasa Corporation is Hindustan Construction Company (64.99%), Avantha Group (16.25%), Venkateshwara Hatcheries (12.8%), and individual investor Vinay Vithal Maniar (6%).[12]

Current progess

As of 2011 four hotels and a city centre have been completed.[15]

Plans

While some sections are complete, construction of Lavasa will not be finished before 2020.[16] Total population when complete will be an estimated 200,000.[17] When complete it will comprise four or five towns built on seven hills.[18] The first of these, Dasve, was supposed to be complete by 2010.[9] Mugaon, the second, may be finished by 2013.[16]

Sports facilities

Planned sports facilities include a Nick Faldo-designed golf course, a Manchester City-branded football academy, and a Hockey Australia hockey academy.[19]

Theme park

A theme park over 65 acres (0.26 km^2) in size is planned.[20]

University

Lavasa will host at least one university.[11] Initially Lavasa had a partnership with Oxford University,[21] [22] but by 2010 this British institution had removed itself from the project.[17] As of 2010, Symbiosis International University is planning to set up a new liberal arts college campus at the township.[23]

Other facilities

A 200 acre health and wellness center[21] and an herb plantation[24] are also planned.

CSR

Lavasa Corporation has initiated a number of CSR programs that target Lavasa and its environs. These include providing vocational training and employment opportunities for locals already living near Lavasa and hosting Diwali festivities for primary school students from nearby villages.[25]

Education

Lavasa will offer a range of educational choices to its residents as well as those who simply come to study.

Hospitality extension campus

Ecole Hoteliere Lavasa is a hospitality management college in Lavasa and an overseas extension of the École hôtelière de Lausanne of Switzerland.[26]

School

A primary, middle, and high school, Le Mont High allows children of Lavasa residents to be educated inside the city's walls.[27]

Future university

Lavasa will host at least one university.[11] Initially Lavasa had a partnership with Oxford University,[21] [22] but by 2010 this British institution had removed itself from the project.[17] As of 2010, Symbiosis International University is planning to set up a new liberal arts college campus at the township.[28]

Awards and recognition

Lavasa has won several awards for its plans and designs. In 2005 Dasve village in Lavasa won awards from the Congress for the New Urbanism and the American Society of Landscape Architects.[29] In 2009 the St. Louis chapter of the American Society of Landscape Architects recognized the Lavasa landscape master plan with a merit award.[30]

Controversies

Lavasa is controversial for many reasons.

Bribes for loans

Lavasa Corporation is accused of benefiting from loans obtained through bribery.[7]

Environmental damage

An Indian Ministry of Environment and Forests team investigating Lavasa concluded the city has caused environmental damage.[2] [31]

Quarrying

While Lavasa has stone crushing permits, its operations have been described as "hill cutting" and "quarrying" by the Indian Ministry of Environment and Forests, and the environmental impact of these activities has been under investigation.[32]

Land acquisition

Some of the land used for Lavasa may have been obtained illegally.

Failure to pay State

A report by the Maharashtra environment department claims 600 hectares (6.0 km^2) of land bought by Lavasa Corporation was purchased from farmers who had been granted it by the Indian State. Because of the way in which the farmers obtained the land, three fourths of the purchase price should have been paid to the State. The report states Lavasa Corporation only paid 2%. It also alleges 141 hectares (1.41 km^2) of Lavasa were leased for far less than actual value by the Maharashtra Krishna Valley Development Corporation and that Lavasa bought 98 hectares (0.98 km^2) of land without license.[12]

Forced removal of landowners

Some say land Lavasa is being built on was obtained through coercion.[6] Lavasa Corporation denies this.[15]

Nepotism

Sharad Pawar, an Indian politician born in the state of Maharashtra, may have demanded compensation for allowing Lavasa to be constructed. When Lavasa Corporation was receiving necessary clearances from the government of Maharashtra, relatives of Pawar had part-ownership of the company developing the project.[33] A daughter and son-in-law of Pawar had more than 20% ownership between 2002 and 2004 and later sold their stakes.[15] A nephew of his was chairman of Maharashtra Krishna Valley Development Corporation (MKVDC) when the MKVDC signed off on lease agreements for Lavasa and allowed it to store water and build dams.[33]

Use of water resources

Lavasa will use the same water resources that currently supply Pune, and it is claimed that this will cause a supply shortage.[24] A 2011 report by the Expert Appraisal Committee of the Union Environment Ministry concluded there will be no impact on Pune's water supply, however.[34]

Orders to Halt Construction

Construction on Lavasa has been suspended due to allegations made by the Indian Ministry of Environment and Forests.[8] This Ministry contends Lavasa did not obtain necessary central government approval obtaining clearance only from the state of Maharashtra instead.[16]

In late 2010 it ordered Lavasa Corporation to halt further construction.[8]

References

[1] http://www.lavasa.com

[2] India rules hill city 'illegal' (http://www.bbc.co.uk/news/world-south-asia-12223024) bbc.co.uk, 19 January 2011 Last updated at 00:57 ET

[3] UPDATE 1-India ministry sets terms to consider Lavasa approval (http://www.reuters.com/article/2011/01/18/lavasa-order-idUSSGE70H0A320110118) reuters.com, Tue Jan 18, 2011 7:58am EST

[4] Biomimicry: Architecture That Imitates Life (http://harvardmagazine.com/2009/09/architecture-imitates-life) harvardmagazine.com, September–October 2009

[5] Lavasa Corporation Ltd: India's First Planned Hill City Deploys Portal Solution to Empower Employees; Increases Collaboration and Efficiency (http://www.microsoft.com/Canada/casestudies/Case_Study_Detail.aspx?casestudyid=4000007468) microsoft.com, 5/31/2010

[6] Howl of the hills (http://www.downtoearth.org.in/node/4987) downtoearth.org.in, by Nidhi Jamwal, Sep 15, 2008

[7] "Suzlon, Lavasa among named in India bribery scam-reports" (http://www.reuters.com/article/idUKSGE6AO03620101125). *Reuters*. November 25, 2010. .

[8] "Environment Ministry asks Lavasa to halt project" (http://economictimes.indiatimes.com/news/news-by-industry/indl-goods-/-svs/construction/Environment-Ministry-asks-Lavasa-to-halt-project/articleshow/6994372.cms). *The Economic Times*. .

[9] India's newest hill station builds for the future (http://www.google.com/hostednews/afp/article/ALeqM5iK6NTbE21WwVHWmJI-4JwBO3LUsg) AFP news hosted by google.com, Sep 15, 2009
[10] HCC's Lavasa court hearing adjourned for 6 weeks (http://www.reuters.com/article/2011/01/27/lavasa-hearing-idUSSGE70Q0AK20110127) reuters.com, Thu Jan 27, 2011 7:31am EST
[11] A Stop in India's Lavasa (http://blogs.forbes.com/timferguson/2010/12/20/a-stop-in-indias-lavasa/) forbes.com, Dec. 20 2010
[12] "SHOWCAUSE NOTICE: JAIRAM ORDERS WORK TO STOP - Lavasa lands in trouble for flouting green laws" (http://epaper.timesofindia.com/Default/Layout/Includes/ETNEW/ArtWin.asp?From=Archive&Source=Page&Skin=ETNEW&BaseHref=ETD/2010/11/27&ViewMode=HTML&GZ=T&PageLabel=1&EntityId=Ar00100&AppName=1). *The Economic Times*. November 27, 2010. . Retrieved February 09, 2011.
[13] "Lavasa IPO may face delay: HCC Chairman" (http://www.business-standard.com/india/news/lavasa-ipo-may-face-delay-hcc-chairman/118365/on). *Business Standard*. .
[14] Pawars no strangers to big land projects (http://www.dnaindia.com/mumbai/report_pawars-no-strangers-to-big-land-projects_1194953-all)
[15] Byatnal, Amruta (October 31, 2010). "Symbolic of luxury, Lavasa is built on irregularities" (http://www.thehindu.com/news/national/article859868.ece). *The Hindu* (Pune). . Retrieved February 21, 2011.
[16] The hills are alive with the sound of controversy (http://www.business-standard.com/india/news/the-hillsalivethe-soundcontroversy/407277/) business-standard.com, September 7, 2010, 1:49 IST
[17] Thekaekara, Tarsh. "The great urban juggernaut" (http://www.newint.org/columns/currents/2010/05/01/lavasa-indias-first-private-city/). New Internationalist. . Retrieved February 9, 2011.
[18] For four towns, see Discover Lavasa: Master Plan; A vision becomes reality (http://www.lavasa.com/high/master_plan.aspx) Lavasa Official Site

- For five towns, see Lavasa: Life in Full (http://www.lavasa.com/high/files/Lavasa_EBrochure.pdf) Lavasa Official Site (**Page 4**)
- For seven hills, see Lavasa Official Site (**page 2**)

[19] For the golf course, see Faldo signs 'stunning' Indian course design project (http://www.nickfaldo.com/archive/news/story/80) nickfaldo.com, 8th June 2009

- For the football academy, see "Blues to open academy in India" (http://www.mcfc.co.uk/News/Club-news/2010/May/Blues-to-open-academy-in-India). *mcfc.co.uk*. 28 May 2010. . Retrieved 28 May 2010-05-28.
- For the hockey academy, see "Hockey Australia recognised as sport world leader following unique partnership" (http://www.hockey.org.au/index.php?id=90&tx_ttnews=298&tx_ttnews=18&cHash=972ef54243). .

[20] http://www.financialexpress.com/news/lavasa-to-host-spaceworld-asias-first-space-edutainment-centre-at-rs-400-crore-investment/413522/
[21] A modern township nestled in the Sahyadris (http://www.thehindubusinessline.in/2008/06/22/stories/2008062251270100.htm) thehindubusinessline.in, 2008/06/22
[22] Oxford dragged into 'land-grab' scam (http://www.hindu.com/2009/08/31/stories/2009083161211800.htm) thehindu.com, by Hasan Suroor. Monday, Aug 31, 2009
[23] Symbiosis to start liberal arts institute (http://www.indianexpress.com/news/symbiosis-to-start-liberal-arts-institute/586682/) indianexpress.com,Thu Mar 04 2010
[24] Water being diverted to Lavasa, alleges Patkar (http://www.expressindia.com/latest-news/water-being-diverted-to-lavasa-alleges-patkar/486570/) expressindia.com, Jul 08, 2009 at 0511 hrs IST
[25] For vocational training, see Lavasa empowers youth from 18 villages (http://www.indiainfoline.com/Markets/News/Lavasa-empowers-youth-from-18-villages/4966532186) indiainfoline.com, Oct 19, 2010

- For children's Diwali festivities, see Lavasa Corporation sparkled hearts of 350 Children this Diwali (http://www.indiainfoline.com/Markets/News/Lavasa-Corporation-sparkled-hearts-of-350-Children-this-Diwali/4977101129) indiainfoline.com, Oct 28, 2010

[26] Ecole Hoteliere Lavasa (http://www.ecolehotelierelavasa.com/) ecolehotelierelavasa.com
[27] From the Head of School's Desk (http://www.le-mont-high.org/about-message-head-of-school.htm) Le Mont High official site, retrieved: 31.8.2011
[28] Symbiosis to start liberal arts institute (http://www.indianexpress.com/news/symbiosis-to-start-liberal-arts-institute/586682/) indianexpress.com,Thu Mar 04 2010
[29] For Congress for the New Urbanism award, see Charter Awards: Past Winners (http://www.cnu.org/awards/winners) Congress for the New Urbanism Official Website

- For American Society of Landscape Architects award, see ASLA Announces 2005 Professional Awards (http://www.asla.org/newsreleasedetails.aspx?id=7770&terms=Lavasa) American Society of Landscape Architects Official Website, 7/11/2005

[30] The 2009 St. Louis Chapter of ASLA Award Winners (http://www.stlouisasla.org/2009-award-winners/) stlouisasla.org
[31] "Maharashtra govt gets MoEF order for action against Lavasa" (http://www.business-standard.com/india/news/maharashtra-govt-gets-moef-order-for-action-against-lavasa/438696/). *Business Standard*. Jun 11, 2011. .
[32] Byatnal, Amruta (Feb 18, 2011). "Lavasa under scanner for quarrying" (http://www.hindu.com/2011/02/18/stories/2011021852450500.htm). *The Hindu* (Chennai, India). .

[33] Vaidya, Abhay (Dec 26, 2010). "Lavasa trail reinforces the Sharad Pawar connection & IPL link" (http://www.dnaindia.com/mumbai/special_lavasa-trail-reinforces-the-sharad-pawar-connection-and-ipl-link_1486204). Pune: DNA. . Retrieved February 21, 2011.
[34] Lavasa showed higher FSI : Jairam Ramesh's MoEF Committee (http://rtn.asia/451_lavasa-showed-higher-fsi-jairam-rameshs-moef) rtn.asia, 05/24/2011

Political corruption

Political corruption is the use of legislated powers by government officials for illegitimate private gain. Misuse of government power for other purposes, such as repression of political opponents and general police brutality, is not considered political corruption. Neither are illegal acts by private persons or corporations not directly involved with the government. An illegal act by an officeholder constitutes political corruption only if the act is directly related to their official duties.

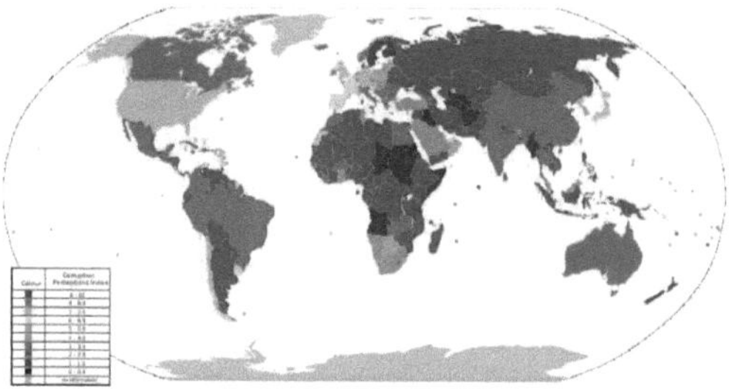

World map of the 2010 Corruption Perceptions Index by Transparency International, which measures "the degree to which corruption is perceived to exist among public officials and politicians". High numbers (blue) indicate less perception of corruption, whereas lower numbers (red) an indicate higher perception of corruption.

Forms of corruption vary, but include bribery, extortion, cronyism, nepotism, patronage, graft, and embezzlement. While corruption may facilitate criminal enterprise such as drug trafficking, money laundering, and human trafficking, it is not restricted to these activities.

The activities that constitute illegal corruption differ depending on the country or jurisdiction. For instance, certain political funding practices that are legal in one place may be illegal in another. In some cases, government officials have broad or poorly defined powers, which make it difficult to distinguish between legal and illegal actions. Worldwide, bribery alone is estimated to involve over 1 trillion US dollars annually.[1] A state of unrestrained political corruption is known as a kleptocracy, literally meaning "rule by thieves".

Effects

Effects on politics, administration, and institutions

Detail from *Corrupt Legislation* (1896) by Elihu Vedder. Library of Congress Thomas Jefferson Building, Washington, D.C.

Corruption poses a serious development challenge. In the political realm, it undermines democracy and good governance by flouting or even subverting formal processes. Corruption in elections and in legislative bodies reduces accountability and distorts representation in policymaking; corruption in the judiciary compromises the rule of law; and corruption in public administration results in the inefficient provision of services. It violates a basic principle of republicanism regarding the centrality of civic virtue. More generally, corruption erodes the institutional capacity of government as procedures are disregarded, resources are siphoned off, and public offices are bought and sold. At the same time, corruption undermines the legitimacy of government and such democratic values as trust and tolerance.

Economic effects

Corruption undermines economic development by generating considerable distortions and inefficiency. In the private sector, corruption increases the cost of business through the price of illicit payments themselves, the management cost of negotiating with officials, and the risk of breached agreements or detection. Although some claim corruption reduces costs by cutting bureaucracy, the availability of bribes can also induce officials to contrive new rules and delays. Openly removing costly and lengthy regulations are better than covertly allowing them to be bypassed by using bribes. Where corruption inflates the cost of business, it also distorts the playing field, shielding firms with connections from competition and thereby sustaining inefficient firms.[2]

Corruption also generates economic distortions in the public sector by diverting public investment into capital projects where bribes and kickbacks are more plentiful. Officials may increase the technical complexity of public sector projects to conceal or pave the way for such dealings, thus further distorting investment. Corruption also lowers compliance with construction, environmental, or other regulations, reduces the quality of government services and infrastructure, and increases budgetary pressures on government.

Economists argue that one of the factors behind the differing economic development in Africa and Asia is that in Africa, corruption has primarily taken the form of rent extraction with the resulting financial capital moved overseas rather than invested at home (hence the stereotypical, but often accurate, image of African dictators having Swiss bank accounts). In Nigeria, for example, more than $400 billion was stolen from the treasury by Nigeria's leaders between 1960 and 1999.[3] University of Massachusetts researchers estimated that from 1970 to 1996, capital flight from 30 sub-Saharan countries totaled $187bn, exceeding those nations' external debts.[4] (The results, expressed in retarded or suppressed development, have been modeled in theory by economist Mancur Olson.) In the case of Africa, one of the factors for this behavior was political instability, and the fact that new governments often confiscated previous government's corruptly-obtained assets. This encouraged officials to stash their wealth abroad, out of reach of any future expropriation. In contrast, Asian administrations such as Suharto's New Order often took a cut on business transactions or provided conditions for development, through infrastructure investment, law and order, etc.

Environmental and social effects

Corruption facilitates environmental destruction. Corrupt countries may formally have legislation to protect the environment, it cannot be enforced if officials can easily be bribed. The same applies to social rights worker protection, unionization prevention, and child labor. Violation of these laws rights enables corrupt countries to gain illegitimate economic advantage in the international market.

The Nobel Prize-winning economist Amartya Sen has observed that "there is no such thing as an apolitical food problem." While drought and other naturally occurring events may trigger famine conditions, it is government action or inaction that determines its severity, and often even whether or not a famine will occur. Governments with strong tendencies towards kleptocracy can undermine food security even when harvests are good. Officials often steal state property. In Bihar, India, more than 80% of the subsidized food aid to poor is stolen by corrupt officials.[5] Similarly, food aid is often robbed at gunpoint by governments, criminals, and warlords alike, and sold for a profit. The 20th century is full of many examples of governments undermining the food security of their own nations – sometimes intentionally.[6]

Effects on Humanitarian Aid

The scale of humanitarian aid to the poor and unstable regions of the world grows, but it is highly vulnerable to corruption, with food aid, construction and other highly valued assistance as the most at risk.[7] Food aid can be directly and physically diverted from its intended destination, or indirectly through the manipulation of assessments, targeting, registration and distributions to favour certain groups or individuals.[7] Elsewhere, in construction and shelter, there are numerous opportunities for diversion and profit through substandard workmanship, kickbacks for contracts and favouritism in the provision of valuable shelter material.[7] Thus while humanitarian aid agencies are usually most concerned about aid being diverted by including too many, recipients themselves are most concerned about exclusion.[7] Access to aid may be limited to those with connections, to those who pay bribes or are forced to give sexual favours.[7] Equally, those able to do so may manipulate statistics to inflate the number beneficiaries and syphon of the additional assistance.[7]

Other areas: health, public safety, education, trade unions, etc.

Corruption is not specific to poor, developing, or transition countries. In western countries, there have been cases of bribery and other forms of corruption in all possible fields: under-the-table payments made to reputed surgeons by patients willing to be on top of the list of forthcoming surgeries,[8] bribes paid by suppliers to the automotive industry in order to sell poor quality connectors used for instance in safety equipment such as airbags, bribes paid by suppliers to manufacturers of defibrillators (to sell poor quality capacitors), contributions paid by wealthy parents to the "social and culture fund" of a prestigious university in exchange for it to accept their children, bribes paid to obtain diplomas, financial and other advantages granted to unionists by members of the executive board of a car manufacturer in exchange for employer-friendly positions and votes, etc. Examples are endless. These various manifestations of corruption can ultimately present a danger for the public health; they can discredit certain essential institutions or social relationships.

Corruption can also affect the various components of sports activities (referees, players, medical and laboratory staff involved in anti-doping controls, members of national sport federation and international committees deciding about the allocation of contracts and competition places).

There have also been cases against (members of) various types of non-profit and non-government organisations, as well as religious organisations.

Ultimately, the distinction between public and private sector corruption sometimes appears rather artificial and national anti-corruption initiatives may need to avoid legal and other loopholes in the coverage of the instruments.

Types

Bribery

A bribe is a payment given personally to a government official in exchange of his use of official powers. Bribery requires two participants: one to give the bribe, and one to take it. Either may initiate the corrupt offering; for example, a customs official may demand bribes to let through allowed (or disallowed) goods, or a smuggler might offer bribes to gain passage. In some countries the culture of corruption extends to every aspect of public life, making it extremely difficult for individuals to stay in business without resorting to bribes. Bribes may be demanded in order for an official to do something he is already paid to do. They may also be demanded in order to bypass laws and regulations. In addition to using bribery for private financial gain, they are also used to intentionally and maliciously cause harm to another (i.e. no financial incentive). In some developing nations, up to half of the population has paid bribes during the past 12 months.[9]

In recent years, efforts have been made by the international community to encourage countries to dissociate and incriminate as separate offences, active and passive bribery. Active bribery can be defined for instance as *the promising, offering or giving by any person, directly or indirectly, of any undue advantage [to any public official], for himself or herself or for anyone else, for him or her to act or refrain from acting in the exercise of his or her functions.* (article 2 of the Criminal Law Convention on Corruption (ETS 173) [10] of the Council of Europe). Passive bribery can be defined as *the request or receipt [by any public official], directly or indirectly, of any undue advantage, for himself or herself or for anyone else, or the acceptance of an offer or a promise of such an advantage, to act or refrain from acting in the exercise of his or her functions* (article 3 of the Criminal Law Convention on Corruption (ETS 173) [10]). The reason for this dissociation is to make the early steps (offering, promising, requesting an advantage) of a corrupt deal already an offence and, thus, to give a clear signal (from a criminal policy point of view) that bribery is not acceptable. Furthermore, such a dissociation makes the prosecution of bribery offences easier since it can be very difficult to prove that two parties (the bribe-giver and the bribe-taker) have formally agreed upon a corrupt deal. In addition, there is often no such formal deal but only a mutual understanding, for instance when it is common knowledge in a municipality that to obtain a building permit one has to pay a "fee" to the decision maker to obtain a favourable decision. A working definition of corruption is also provided as follows in article 3 of the Civil Law Convention on Corruption (ETS 174) [11]: *For the purpose of this Convention, "corruption" means requesting, offering, giving or accepting, directly or indirectly, a bribe or any other undue advantage or prospect thereof, which distorts the proper performance of any duty or behavior required of the recipient of the bribe, the undue advantage or the prospect thereof.*

Trading in influence

Trading in influence, or influence peddling in certain countries, refers to the situation where a person is selling his/her influence over the decision process involving a third party (person or institution). The difference with bribery is that this is a tri-lateral relation. From a legal point of view, the role of the third party (who is the target of the influence) does not really matter although he/she can be an accessory in some instances. It can be difficult to make a distinction between this form of corruption and certain forms of extreme and poorly regulated lobbying where for instance law- or decision-makers can freely "sell" their vote, decision power or influence to those lobbyists who offer the highest retribution, including where for instance the latter act on behalf of powerful clients such as industrial groups who want to avoid the passing of certain environmental, social, or other regulations perceived as too stringent, etc. Where lobbying is (sufficiently) regulated, it becomes possible to provide for a distinctive criteria and to consider that trading in influence involves the use of "improper influence", as in article 12 of the Criminal Law

Convention on Corruption (ETS 173) [10] of the Council of Europe.

Patronage

Patronage refers to favoring supporters, for example with government employment. This may be legitimate, as when a newly elected government changes the top officials in the administration in order to effectively implement its policy. It can be seen as corruption if this means that incompetent persons, as a payment for supporting the regime, are selected before more able ones. In nondemocracies many government officials are often selected for loyalty rather than ability. They may be almost exclusively selected from a particular group (for example, Sunni Arabs in Saddam Hussein's Iraq, the nomenklatura in the Soviet Union, or the Junkers in Imperial Germany) that support the regime in return for such favors. A similar problem can also be seen in Eastern Europe, for example in Romania, where the government is often accused of patronage (when a new government comes to power it rapidly changes most of the officials in the public sector).

Nepotism and cronyism

Favoring relatives (nepotism) or personal friends (cronyism) of an official is a form of illegitimate private gain. This may be combined with bribery, for example demanding that a business should employ a relative of an official controlling regulations affecting the business. The most extreme example is when the entire state is inherited, as in North Korea or Syria. A milder form of cronyism is a "Good ol' boy network", in which appointees to official positions are selected only from a closed and exclusive social network – such as the alumni of particular universities – instead of appointing the most competent candidate.

Seeking to harm enemies becomes corruption when official powers are illegitimately used as means to this end. For example, trumped-up charges are often brought up against journalists or writers who bring up politically sensitive issues, such as a politician's acceptance of bribes.

Electoral fraud

Electoral fraud is illegal interference with the process of an election. Acts of fraud affect vote counts to bring about an election result, whether by increasing the vote share of the favored candidate, depressing the vote share of the rival candidates, or both. Also called **voter fraud**, the mechanisms involved include illegal voter registration, intimidation at polls, and improper vote counting.

Embezzlement

Embezzlement is outright theft of entrusted funds. It is a misappropriation of property.

Another common type of embezzlement is that of entrusted government resources; for example, when a director of a public enterprise employs company workers to build or renovate his own house.

Kickbacks

A kickback is an official's share of misappropriated funds allocated from his or her organization to an organization involved in corrupt bidding. For example, suppose that a politician is in charge of choosing how to spend some public funds. He can give a contract to a company that is not the best bidder, or allocate more than they deserve. In this case, the company benefits, and in exchange for betraying the public, the official receives a kickback payment, which is a portion of the sum the company received. This sum itself may be all or a portion of the difference between the actual (inflated) payment to the company and the (lower) market-based price that would have been paid had the bidding been competitive. Kickbacks are not limited to government officials; any situation in which people are entrusted to spend funds that do not belong to them are susceptible to this kind of corruption. Kickbacks are also common in the pharmaceutical industry, as many doctors and physicians receive pay in return for added promotion and prescription of the drug these pharmaceutical companies are marketing.

Unholy alliance

An unholy alliance is a coalition among seemingly antagonistic groups, especially if one is religious,[12] for ad hoc or hidden gain. Like patronage, unholy alliances are not necessarily illegal, but unlike patronage, by its deceptive nature and often great financial resources, an unholy alliance can be much more dangerous to the public interest. An early, well-known use of the term was by Theodore Roosevelt (TR):

> "To destroy this invisible Government, to dissolve the **unholy alliance** between corrupt business and corrupt politics is the first task of the statesmanship of the day." – 1912 Progressive Party Platform, attributed to TR[13] and quoted again in his autobiography[14] where he connects trusts and monopolies (sugar interests, Standard Oil, etc.) to Woodrow Wilson, Howard Taft, and consequently both major political parties.

Involvement in organized crime

An illustrative example of official involvement in organized crime can be found from 1920s and 1930s Shanghai, where Huang Jinrong was a police chief in the French concession, while simultaneously being a gang boss and co-operating with Du Yuesheng, the local gang ringleader. The relationship kept the flow of profits from the gang's gambling dens, prostitution, and protection rackets undisturbed.

The United States accused Manuel Noriega's government in Panama of being a "narcokleptocracy", a corrupt government profiting on illegal drug trade. Later the U.S. invaded Panama and captured Noriega.

Conditions favorable for corruption

It is argued that the following conditions are favorable for corruption:

- Information deficits
  - Lacking freedom of information legislation. For example: The Indian Right to Information Act 2005 is perceived to have "already engendered mass movements in the country that is bringing the lethargic, often corrupt bureaucracy to its knees and changing power equations completely."[15]
  - Lack of investigative reporting in the local media.
  - Contempt for or negligence of exercising freedom of speech and freedom of the press.
  - Weak accounting practices, including lack of timely financial management.
  - Lack of measurement of corruption. For example, using regular surveys of households and businesses in order to quantify the degree of perception of corruption in different parts of a nation or in different government institutions may increase awareness of corruption and create pressure to combat it. This will also enable an evaluation of the officials who are fighting corruption and the methods used.
  - Tax havens which tax their own citizens and companies but not those from other nations and refuse to disclose information necessary for foreign taxation. This enables large scale political corruption in the foreign nations.[16]
- Lacking control of the government.
  - Lacking civic society and non-governmental organizations which monitor the government.
  - An individual voter may have a rational ignorance regarding politics, especially in nationwide elections, since each vote has little weight.
  - Weak civil service, and slow pace of reform.
  - Weak rule of law.
  - Weak legal profession.
  - Weak judicial independence.
  - Lacking protection of whistleblowers.
    - Government Accountability Project

  - Lack of benchmarking, that is continual detailed evaluation of procedures and comparison to others who do similar things, in the same government or others, in particular comparison to those who do the best work. The Peruvian organization Ciudadanos al Dia has started to measure and compare transparency, costs, and efficiency in different government departments in Peru. It annually awards the best practices which has received widespread media attention. This has created competition among government agencies in order to improve.[17]
- Opportunities and incentives
  - Individual officials routinely handle cash, instead of handling payments by giro or on a separate cash desk—illegitimate withdrawals from supervised bank accounts are much more difficult to conceal.
  - Public funds are centralized rather than distributed. For example, if $1,000 is embezzled from a local agency that has $2,000 funds, it is easier to notice than from a national agency with $2,000,000 funds. See the principle of subsidiarity.
  - Large, unsupervised public investments.
  - Sale of state-owned property and privatization.
  - Poorly-paid government officials.
  - Government licenses needed to conduct business, e.g., import licenses, encourage bribing and kickbacks.
  - Long-time work in the same position may create relationships inside and outside the government which encourage and help conceal corruption and favoritism. Rotating government officials to different positions and geographic areas may help prevent this; for instance certain high rank officials in French government services (e.g. treasurer-paymasters general) must rotate every few years.
  - Costly political campaigns, with expenses exceeding normal sources of political funding, especially when funded with taxpayer money.
  - Less interaction with officials reduces the opportunities for corruption. For example, using the Internet for sending in required information, like applications and tax forms, and then processing this with automated computer systems. This may also speed up the processing and reduce unintentional human errors. See e-Government.
  - A windfall from exporting abundant natural resources may encourage corruption.[18] *(See Resource curse)*
  - War and other forms of conflict correlate with a breakdown of public security.
- Social conditions
  - Self-interested closed cliques and "old boy networks".
  - Family-, and clan-centered social structure, with a tradition of nepotism/favouritism being acceptable.
  - A gift economy, such as the Soviet blat system, emerges in a Communist centrally planned economy.
  - Lacking literacy and education among the population.
  - Frequent discrimination and bullying among the population.
  - Tribal solidarity, giving benefits to certain ethnic groups

In the Indian political system, for example, it has become usual that the leadership of national and regional parties are passed from generation to generation[19] [20] creating a system in which a family holds the center of power. Some examples are most of the Dravidian parties of south India and also the Congress party, which is one of the two major political parties in India.

Size of public sector

Extensive and diverse public spending is, in itself, inherently at risk of cronyism, kickbacks, and embezzlement. Complicated regulations and arbitrary, unsupervised official conduct exacerbate the problem. This is one argument for privatization and deregulation. Opponents of privatization see the argument as ideological. The argument that corruption necessarily follows from the opportunity is weakened by the existence of countries with low to non-existent corruption but large public sectors, like the Nordic countries.[21] However, these countries score high on the Ease of Doing Business Index, due to good and often simple regulations, and have rule of law firmly established. Therefore, due to their lack of corruption in the first place, they can run large public sectors without inducing political corruption.

Like other governmental economic activities, also privatization, such as in the sale of government-owned property, is particularly at the risk of cronyism. Privatizations in Russia, Latin America, and East Germany were accompanied by large scale corruption during the sale of the state owned companies. Those with political connections unfairly gained large wealth, which has discredited privatization in these regions. While media have reported widely the grand corruption that accompanied the sales, studies have argued that in addition to increased operating efficiency, daily petty corruption is, or would be, larger without privatization, and that corruption is more prevalent in non-privatized sectors. Furthermore, there is evidence to suggest that extralegal and unofficial activities are more prevalent in countries that privatized less.[22]

There is the counter point, however, that oligarchy industries can be quite corrupt ("competition" like collusive price-fixing, pressuring dependent businesses, etc.), and only by having a portion of the market owned by someone *other than* that oligarchy, i.e. public sector, can keep them in line (if the public sector gas company is making money & selling gas for 1/2 of the price of the private sector companies... the private sector companies won't be able to simultaneously gouge to that degree & keep their customers: the competition keeps them in line). Private sector corruption can increase the poverty/helplessness of the population, so it can affect government corruption, in the long-term.

In the European Union, the principle of subsidiarity is applied: a government service should be provided by the lowest, most local authority that can competently provide it. An effect is that distribution of funds into multiple instances discourages embezzlement, because even small sums missing will be noticed. In contrast, in a centralized authority, even minute proportions of public funds can be large sums of money.

Governmental corruption

If the highest echelons of the governments also take advantage from corruption or embezzlement from the state's treasury, it is sometimes referred with the neologism *kleptocracy*. Members of the government can take advantage of the natural resources (e.g., diamonds and oil in a few prominent cases) or state-owned productive industries. A number of corrupt governments have enriched themselves via foreign aid, which is often spent on showy buildings and armaments.

A corrupt dictatorship typically results in many years of general hardship and suffering for the vast majority of citizens as civil society and the rule of law disintegrate. In addition, corrupt dictators routinely ignore economic and social problems in their quest to amass ever more wealth and power.

The classic case of a corrupt, exploitive dictator often given is the regime of Marshal Mobutu Sese Seko, who ruled the Democratic Republic of the Congo (which he renamed Zaire) from 1965 to 1997. It is said that usage of the term kleptocracy gained popularity largely in response to a need to accurately describe Mobutu's regime. Another classic case is Nigeria, especially under the rule of General Sani Abacha who was *de facto* president of Nigeria from 1993 until his death in 1998. He is reputed to have stolen some US$3–4 billion. He and his relatives are often mentioned in Nigerian 419 letter scams claiming to offer vast fortunes for "help" in laundering his stolen "fortunes", which in reality turn out not to exist.[23] More than $400 billion was stolen from the treasury by Nigeria's leaders between 1960 and 1999.[24]

More recently, articles in various financial periodicals, most notably *Forbes* magazine, have pointed to Fidel Castro, General Secretary of the Republic of Cuba since 1959, of likely being the beneficiary of up to $900 million, based on "his control" of state-owned companies.[25] Opponents of his regime claim that he has used money amassed through weapons sales, narcotics, international loans, and confiscation of private property to enrich himself and his political cronies who hold his dictatorship together, and that the $900 million published by Forbes is merely a portion of his assets, although that needs to be proven.[26]

Judiciary corruption

It is one of the most dangerous type of corruption because people are loosing their trust into justice. Budget of the judiciary in many transitional and developing countries is almost completely controlled by the executive. The latter undermines the separation of powers, as it creates a critical financial dependence of the judiciary. The proper national wealth distribution including the government spending on the judiciary is subject of the constitutional economics. It is important to distinguish between the two methods of corruption of the judiciary: the state (through budget planning and various privileges), and the private.[27]

Fighting corruption

Mobile telecommunications and radio broadcasting help to fight corruption, especially in developing regions like Africa,[28] where other forms of communications are limited.

In the 1990s, initiatives were taken at an international level (in particular by the European Community, the Council of Europe, the OECD) to put a ban on corruption: in 1996, the Committee of Ministers of the Council of Europe,[29] for instance, adopted a comprehensive Programme of Action against Corruption and, subsequently, issued a series of anti-corruption standard-setting instruments:

- the Criminal Law Convention on Corruption (ETS 173);[30]
- the Civil Law Convention on Corruption (ETS 174);[31]
- the Additional Protocol to the Criminal Law Convention on Corruption (ETS 191);[32]
- the Twenty Guiding Principles for the Fight against Corruption (Resolution (97) 24);[33]
- the Recommendation on Codes of Conduct for Public Officials (Recommendation No. R (2000) 10);[34]
- the Recommendation on Common Rules against Corruption in the Funding of Political Parties and Electoral Campaigns (Rec(2003)4)[35]

The purpose of these instruments was to address the various forms of corruption (involving the public sector, the private sector, the financing of political activities, etc.) whether they had a strictly domestic or also a transnational dimension. To monitor the implementation at national level of the requirements and principles provided in those texts, a monitoring mechanism – the Group of States Against Corruption (also known as GRECO) was created.

Further conventions were adopted at the regional level under the aegis of the Organization of American States (OAS or OEA), the African Union, and in 2003, at the universal level under that of the United Nations.

Campaign contributions

In the political arena, it is difficult to prove corruption. For this reason, there are often unproven rumors about many politicians, sometimes part of a smear campaign.

Politicians are placed in apparently compromising positions because of their need to solicit financial contributions for their campaign finance. If they then appear to be acting in the interests of those parties that funded them, it could be considered corruption. Though donations may be coincidental, the question asked is, why are they funding politicians at all, if they get nothing for their money.

Laws regulating campaign finance in the United States require that all contributions and their use should be publicly disclosed. Many companies, especially larger ones, fund both the Democratic and Republican parties. Certain countries, such as France, ban altogether the corporate funding of political parties. Because of the possible circumvention of this ban with respect to the funding of political campaigns, France also imposes maximum spending caps on campaigning; candidates that have exceeded those limits, or that have handed in misleading accounting reports, risk having their candidacy ruled invalid, or even being prevented from running in future elections. In addition, the government funds political parties according to their successes in elections.

In some countries, political parties are run solely off subscriptions (membership fees).

Even legal measures such as these have been argued to be legalized corruption, in that they often favor the political status quo. Minor parties and independents often argue that efforts to rein in the influence of contributions do little more than protect the major parties with guaranteed public funding while constraining the possibility of private funding by outsiders. In these instances, officials are legally taking money from the public coffers for their election campaigns to guarantee that they will continue to hold their influential and often well-paid positions.

As indicated above, the Committee of Ministers of the Council of Europe recognised in 1996 the importance of links between corruption and political financing. It adopted in 1837 the Recommendation on Common Rules against Corruption in the Funding of Political Parties and Electoral Campaigns (Rec(2003)4) [36]. This text is quite unique at international levels as it aims i.a. at increasing transparency in the funding of political parties and election campaigns (these two areas are difficult to dissociate since parties are also involved in campaigning and in many countries, parties do not have the monopoly over the presentation of candidates for elections), ensuring a certain level of control over the funding and spending connected with political activities, and making sure infringements are subject to effective, proportionate, and dissuasive sanctions. In the context of its monitoring activities, the Group of States Against Corruption has identified a great variety of possible improvements in those areas (see the country reports adopted under the Third Evaluation Round [37]).

Measuring corruption

Measuring corruption statistically is difficult if not impossible due to the illicit nature of the transaction and imprecise definitions of corruption.[38] While "corruption" indices first appeared in 1995 with the Corruption Perceptions Index, all of these metrics address different proxies for corruption, such as public perceptions of the extent of the problem.[39]

Transparency International, an anti-corruption NGO, pioneered this field with the Corruption Perceptions Index, first released in 1995. This work is often credited with breaking a taboo and forcing the issue of corruption into high level development policy discourse. Transparency International currently publishes three measures, updated annually: a Corruption Perceptions Index(CPI) (based on aggregating third-party polling of public perceptions of how corrupt different countries are); a Global Corruption Barometer (based on a survey of general public attitudes toward and experience of corruption); and a Bribe Payers Index, looking at the willingness of foreign firms to pay bribes. The Corruption Perceptions Index is the best known of these metrics, though it has drawn much criticism[39] [40] [41] and may be declining in influence.[42]

The World Bank collects a range of data on corruption [43], including survey responses from over 100,000 firms worldwide [44] and a set of indicators of governance and institutional quality. [45] Moreover, one of the six dimensions of governance measured by the Worldwide Governance Indicators is Control of Corruption, which is defined as "the extent to which power is exercised for private gain, including both petty and grand forms of corruption, as well as 'capture' of the state by elites and private interests."[46] While the definition itself is fairly precise, the data aggregated into the Worldwide Governance Indicators is based on any available polling: questions range from "is corruption a serious problem?" to measures of public access to information, and not consistent across countries. Despite these weaknesses, the global coverage of these datasets has led to their widespread adoption, most notably by the Millennium Challenge Corporation.[38]

In part in response to these criticisms, a second wave of corruption metrics has been created by Global Integrity, the International Budget Partnership [47], and many lesser known local groups, starting with the Global Integrity Index,[48] first published in 2004. These second wave projects aim not to create awareness, but to create policy change via targeting resources more effectively and creating checklists toward incremental reform. Global Integrity and the International Budget Partnership[49] each dispense with public surveys and instead uses in-country experts to evaluate "the opposite of corruption" – which Global Integrity defines as the public policies that prevent, discourage, or expose corruption.[50] These approaches compliment the first wave, awareness-raising tools by giving governments facing public outcry a checklist which measures concrete steps toward improved governance.[38]

Typical second wave corruption metrics do not offer the worldwide coverage found in first wave projects, and instead focus on localizing information gathered to specific problems and creating deep, "unpackable" content that matches quantitative and qualitative data. Meanwhile, alternative approaches such as the British aid agency's Drivers of Change research skips numbers entirely and favors understanding corruption via political economy analysis of who controls power in a given society.[38]

Institutions dealing with political corruption

- Global Witness, an international NGO established in 1993 that works to break the links between natural resource exploitation, conflict, poverty, corruption, and human rights abuses worldwide
- Group of States Against Corruption, a body established under the Council of Europe to monitor the implementation of instruments adopted by member states to combat political corruption
- Independent Commission Against Corruption (disambiguation)
- Transparency International, a non-governmental organization that monitors and publicizes corporate and political corruption in international development
    - Corruption Perceptions Index, published yearly by Transparency International
- TrustLaw, a service of the Thomson Reuters Foundation is a global hub for free legal assistance and news and information on anti-corruption

See also

- Corruption by country
- Corruption in local government
- John Hoyle
- Independence of the judiciary
- List of politicians in India charged with corruption
- Police corruption
- Political class
- Political machine
- Power elite

Forms or aspects of corruption

- Baksheesh
- Crony capitalism
- Electoral fraud
- Honest services fraud
- Political scandal
- Professional courtesy
- Systemic corruption

Good governance

- Civil society
- Comitology
- Constitutional economics
- Due diligence

Theoretical aspects

- Conflict of interest
- Principal-agent problem
- Rent seeking

Anti-corruption authorities and measures

- FBI[51]
- GEMAP
- The Council of Europe's Group of States Against Corruption (French: Groupe d'Etats contre la corruption) or GRECO, the body and mechanism established to monitor the implementation of:
  - the Criminal Law Convention on Corruption (ETS 173);[52]
  - the Civil Law Convention on Corruption (ETS 174);[53]
  - the Additional Protocol to the Criminal Law Convention on Corruption (ETS 191);[54]
  - the Twenty Guiding Principles for the Fight against Corruption (Resolution (97) 24);[55]
  - the Recommendation on Codes of Conduct for Public Officials (Recommendation No. R (2000) 10);[56]
  - the Recommendation on Common Rules against Corruption in the Funding of Political Parties and Electoral Campaigns (Rec(2003)4)[57]
- Independent Commission Against Corruption
- India Against Corruption[58]
- Inter-American Convention Against Corruption
- Report Public Corruption
- United Nations Convention against Corruption

India Against corruption logo

Corruption in fiction

- *The Financier* (1912), *The Titan* (1914), and *The Stoic* (1947), Theodore Dreiser's *Trilogy of Desire*, based on the life of the notorious transit mogul Charles Tyson Yerkes
- *Mr. Smith Goes to Washington* (Hollywood film 1939)
- *Atlas Shrugged* (1957 novel)
- Henry Adams' novel *Democracy* (1880)
- Carl Hiaasen's novel *Sick Puppy* (1999)
- Much of the Batman comic book series
- *V for Vendetta* comic book series
- The Ghost in the Shell Anime films and series
- *Animal Farm* a novel by George Orwell
- *Training Day* (2001 film)
- *Exit Wounds* (2001 film)
- *American Gangster* (2007 film)
- *Robert Penn Warren's novel All the King's Men* (1946)

References

[1] African corruption 'on the wane' (http://news.bbc.co.uk/1/hi/business/6288400.stm), 10 July 2007, BBC News

[2] Luis Flores Ballesteros, "Corruption and development. Does the "rule of law" factor weigh more than we think?" (http://54pesos.org/2008/11/15/corruption-and-development-does-the-ârule-of-lawâ-factor-weigh-more-than-we-think) 54 Pesos (November 15, 2008). Retrieved April 12, 2011

[3] "Nigeria's corruption busters" (http://www.unodc.org/unodc/en/frontpage/nigerias-corruption-busters.html). Unodc.org. . Retrieved 2009-12-05.

[4] "When the money goes west" (http://www.newstatesman.com/Economy/200503140015). New Statesman. 2005-03-14. . Retrieved 2009-11-05.

[5] "Will Growth Slow Corruption In India?" (http://www.forbes.com/2007/08/15/wipro-tata-corruption-ent-law-cx_kw_0814whartonindia.html). Forbes. 2007-08-15. .

[6] Sheeter, Laura (2007-11-24). "Ukraine remembers famine horror" (http://news.bbc.co.uk/2/hi/europe/7111296.stm). BBC News. . Retrieved 2009-12-05.

[7] Sarah Bailey (2008) Need and greed: corruption risks, perceptions and prevention in humanitarian assistance (http://www.odi.org.uk/resources/details.asp?id=2385&title=corruption-risks-perceptions-prevention-humanitarian-assistance) Overseas Development Institute

[8] Fidelman, Charlie (November 27, 2010). "Cash bribes put patients atop surgery waiting lists" (http://www.vancouversun.com/health/Cash+bribes+patients+atop+surgery+waiting+lists/3893239/story.html). *The Vancouver Sun*. . Retrieved 2011-01-21.

[9] "How common is bribe-paying?" (http://www.transparency.org/news_room/latest_news/press_releases/2005/09_12_2005_barometer_2005). . "...a relatively high proportion of families in a group of Central Eastern European, African, and Latin American countries paid a bribe in the previous twelve months."

[10] http://conventions.coe.int/treaty/Commun/QueVoulezVous.asp?NT=173&CL=ENG

[11] http://conventions.coe.int/Treaty/Commun/QueVoulezVous.asp?NT=174&CM=1&DF=7/18/2008&CL=ENG

[12] "Kentucky's unholy alliance" (http://kpac2.blogspot.com/2007/04/unholy-alliance.html). Kpac2.blogspot.com. 2007-04-05. . Retrieved 2009-12-05.

[13] By Patricia O*Toole Sunday, Jun. 25, 2006 (2006-06-25). "O'Toole, Patricia, "The War of 1912," TIME in Partneship with CNN, Jun. 25, 2006"* (http://www.time.com/time/magazine/article/0,9171,1207791-2,00.html). *Time.com. . Retrieved 2009-12-05.*

[14] "Roosevelt, Theodore. "An Autobiography:" XV. The Peace of Righteousness, Appendix B, New York: Macmillan, 1913" (http://www.bartleby.com/55/15b.html). Bartleby.com. . Retrieved 2009-12-05.

[15] "AsiaMedia :: Right to Information Act India's magic wand against corruption" (http://www.asiamedia.ucla.edu/article.asp?parentid=52046). Asiamedia.ucla.edu. 2006-08-31. . Retrieved 2009-11-05.

[16] Mathiason, Nick (2007-01-21). "Western bankers and lawyers 'rob Africa of $150bn every year" (http://observer.guardian.co.uk/business/story/0,,1994976,00.html). London: Observer.guardian.co.uk. . Retrieved 2009-12-05.

[17] "Why benchmarking works – PSD Blog – World Bank Group" (http://psdblog.worldbank.org/psdblog/2006/08/why_benchmarkin.html#more). Psdblog.worldbank.org. 2006-08-17. . Retrieved 2009-11-05.

[18] Damania, Richard; Bulte, Erwin (July 2003). "Resources for Sale: Corruption, Democracy and the Natural Resource Curse" (http://www.adelaide.edu.au/cies/papers/0320.pdf) (pdf). Centre for International Economic Studies, University of Adelaide. . Retrieved 2010-12-11.

[19] Soutik Biswas (1-18-2011). "Is India sliding into a hereditary monarchy?" (http://www.bbc.co.uk/blogs/thereporters/soutikbiswas/2011/01/is_india_sliding_into_a.html). *BBC*. BBC News. . Retrieved 3 September 2011.

[20] Deo, Manjeet; Kripalani (8-5-2011). "The Gandhi dynasty: Politics as usual" (http://www.rediff.com/news/column/the-gandhi-dynasty-politics-as-usual/20110805.htm). *Rediff*. Rediff News. . Retrieved 3 September 2011.

[21] "Lessons From the North" (http://www.project-syndicate.org/commentary/sachs110). Project Syndicate. . Retrieved 2009-11-05.

[22] Privatization in Competitive Sectors: The Record to Date. Sunita Kikeri and John Nellis. World Bank Policy Research Working Paper 2860, June 2002. Econ.Chula.ac.th (http://www.econ.chula.ac.th/about/member/sothitorn/privatization.pdf) Privatization and Corruption. David Martimort and Stéphane Straub. IDEI.fr (http://idei.fr/doc/conf/veol/straub_martimort.pdf)

[23] Who wants to be a millionaire? – An online collection of Nigerian scam mails

[24] "Nigeria's corruption totals $400 billion" (http://web.archive.org/web/20071211033642/http://web.archive.org/web/20071211033642/http://www.malaysia-today.net/Blog-e/2005/06/nigerias-corruption-totals-400-billion.htm). *Malaysia Today*. June 27, 2005. Archived from the original on 2007-12-11. .

[25] "Fidel Castro net worth rises, according to 'Forbes'" (http://www.usatoday.com/money/2006-05-04-castro_x.htm). Usatoday.Com. 2006-05-04. . Retrieved 2009-11-05.

[26] Shapiro, Ben (2006-08-05). "Ben Shapiro :: Townhall.com :: The death of Fidel Castro" (http://www.townhall.com/columnists/BenShapiro/2006/08/02/the_death_of_fidel_castro). Townhall.com. . Retrieved 2009-11-05.

[27] Barenboim, Peter (October 2009). *Defining the rules*. The European Lawyer.

[28] "Mobile Phones and Radios Combat Corruption in Burundi – Voices from Emerging Markets" (http://voicesfromemergingmarkets.com/?p=19). Voicesfromemergingmarkets.com. 2009-03-12. . Retrieved 2009-11-05.

[29] http://www.coe.int/t/cm/Home_en.asp

[30] http://conventions.coe.int/treaty/Commun/QueVoulezVous.asp?NT=173&CL=ENG

[31] http://conventions.coe.int/Treaty/Commun/QueVoulezVous.asp?NT=174&CM=1&DF=7/18/2008&CL=ENG

[32] http://conventions.coe.int/Treaty/Commun/QueVoulezVous.asp?NT=191&CM=1&DF=7/18/2008&CL=ENG
[33] http://www.coe.int/t/dghl/monitoring/greco/documents/Resolution(97)24_EN.pdf
[34] http://www.coe.int/t/dghl/monitoring/greco/documents/Rec(2000)10_EN.pdf
[35] http://www.coe.int/t/dghl/monitoring/greco/general/Rec(2003)4_EN.pdf
[36] http://www.coe.int/t/dghl/monitoring/greco/general/Rec(2003)4_EN.pdf
[37] http://www.coe.int/t/dghl/monitoring/greco/evaluations/index_en.asp
[38] "A Users' Guide to Measuring Corruption" (http://commons.globalintegrity.org/2008/09/users-guide-to-measuring-corruption.html). Global Integrity. September 5, 2008. . Retrieved 2010-12-11.
[39] Galtung, Fredrik (2006). "Measuring the Immeasurable: Boundaries and Functions of (Macro) Corruption Indices," in Measuring Corruption, Charles Sampford, Arthur Shacklock, Carmel Connors, and Fredrik Galtung, Eds. (Ashgate): 101–130.
[40] Sik, Endre (2002). "The Bad, the Worse and the Worst: Guesstimating the Level of Corruption," in Political Corruption in Transition: A Skeptic's Handbook, Stephen Kotkin and Andras Sajo, Eds. (Budapest: Central European University Press): 91–113.
[41] Arndt, Christiane and Charles Oman (2006). Uses and Abuses of Governance Indicators (Paris: OECD Development Centre).
[42] "Media citing Transparency International" (http://www.google.com/trends?q=Transparency+International&ctab=-1&geo=all&date=all). Google.com. . Retrieved 2009-12-05.
[43] http://www.worldbank.org/wbi/governance/data/
[44] http://www.enterprisesurveys.org/ExploreTopics/?topicid=3
[45] http://go.worldbank.org/IU54LJVHT0
[46] "A Decade of Measuring the Quality of Governance" (http://web.archive.org/web/20080527211855/http://web.archive.org/web/20080527211855/http://info.worldbank.org/governance/wgi2007/pdf/booklet_decade_of_measuring_governance.pdf). The International Bank for Reconstruction and Development, The World Bank. 2007. p. 3. Archived from the original on 2008-05-27. .
[47] http://www.internationalbudget.org/
[48] http://report.globalintegrity.org/
[49] http://www.internationalbudget.org/
[50] "The Global Integrity Report: 2009 Methodology White Paper" (http://report.globalintegrity.org/methodology/whitepaper.cfm). Global Integrity. 2009. . Retrieved 2010-12-11.
[51] http://www.fbi.gov/about-us/investigate/corruption
[52] http://conventions.coe.int/treaty/Commun/QueVoulezVous.asp?NT=173&CL=ENG
[53] http://conventions.coe.int/Treaty/Commun/QueVoulezVous.asp?NT=174&CM=1&DF=7/18/2008&CL=ENG
[54] http://conventions.coe.int/Treaty/Commun/QueVoulezVous.asp?NT=191&CM=1&DF=7/18/2008&CL=ENG
[55] http://www.coe.int/t/dghl/monitoring/greco/documents/Resolution(97)24_EN.pdf
[56] http://www.coe.int/t/dghl/monitoring/greco/documents/Rec(2000)10_EN.pdf
[57] http://www.coe.int/t/dghl/monitoring/greco/general/Rec(2003)4_EN.pdf
[58] http://www.indiaagainstcorruption.org/

Further reading

- Michael W. Collier. (2009) *Political Corruption in the Caribbean Basin: Constructing a Theory to Combat Corruption* excerpt and text search (http://www.amazon.com/dp/0415804922)
- Charles Copeman and Amy McGrath (eds.)(1997), *Corrupt Elections. Ballot Rigging in Australia*, Towerhouse Publications, Kensington, NSW
- Donatella della Porta, and Alberto Vannucci, (1999). *Corrupt Exchanges: Actors, Resources, and Mechanisms of Political Corruption.* New York: Aldine de Gruyter.
- Axel Dreher, Christos Kotsogiannis, Steve McCorriston (2004), Corruption Around the World: Evidence from a Structural Model (http://econwpa.wustl.edu/eps/pe/papers/0406/0406004.pdf).
- Kimberly Ann Elliott, (ed.) (1997) *Corruption and the Global Economy*
- Edward L. Glaeser and Claudia Goldin, (eds.) (2006), *Corruption and Reform: Lessons from America's Economic History* U. of Chicago Press, 386 pp. ISBN 0-226-29957-0.
- Mark Grossman. *Political Corruption in America: An Encyclopedia of Scandals, Power, and Greed* (2 vol. 2008)
- Arnold J. Heidenheimer, Michael Johnston and Victor T. LeVine (eds.) (1989), *Political Corruption: A Handbook* 1017 pages.
- Richard Jensen. (2001) "Democracy, Republicanism and Efficiency: The Values of American Politics, 1885–1930," in Byron Shafer and Anthony Badger, eds, *Contesting Democracy: Substance and Structure in American Political History, 1775–2000* pp 149–180; online edition (http://www.uic.edu/~rjensen/rj0025.htm)

- Michael Johnston, Victor T. LeVine, and Arnold Heidenheimer, eds. (1970) *Political Corruption: Readings in Comparative Analysis*
- Michael Johnston (2005), *Syndromes of Corruption: Wealth, Power, and Democracy*
- Junichi Kawata. (2006) *Comparing Political Corruption And Clientelism* excerpt and text search (http://www.amazon.com/dp/0754643565)
- George C. Kohn (2001). *The New Encyclopedia of American Scandal*
- Johann Graf Lambsdorff (2007), *The Institutional Economics of Corruption and Reform: Theory, Evidence and Policy* (http://www.cambridge.org/catalogue/catalogue.asp?isbn=0521872758) Cambridge University Press
- Amy McGrath, (1994), *The Forging of Votes*, Tower House Publications, Kensington, NSW
- Amy McGrath, (2003), *Frauding of Elections*, Tower House Publications and H.S. Chapman Society, Brighton-le Sands, NSW
- Amy McGrath, (1994), *The Frauding of Votes*, Tower House Publications, Kensington, NSW
- Amy McGrath, (2005), *The Stolen Election, Australia 1987 According to Frank Hardy, Author of Power Without Glory*, Towerhouse Publications and H.S. Chapman Society, Brighton-le Sands, NSW
- John Mukum Mbaku. (1999) *Bureaucratic and Political Corruption in Africa: The Public Choice Perspective*
- Stephen D. Morris. (2009) *Political Corruption in Mexico: The Impact of Democratization*
- Aaron G. Murphy. (2010) *Foreign Corrupt Practices Act: A Practical Resource for Managers and Executives* (http://www.amazon.com/Foreign-Corrupt-Practices-Act-Executives/dp/0470918004/ref=sr_1_2?ie=UTF8&qid=1298574989&sr=8-2)
- Peter John Perry. (2002) *Political Corruption in Australia: A Very Wicked Place?*
- John F. Reynolds. (1988). *Testing Democracy: Electoral Behavior and Progressive Reform in New Jersey, 1880–1920* on corrupt voting methods
- Robert North Roberts. (2001) *Ethics in U.S. Government: An Encyclopedia of Investigations, Scandals, Reforms, and Legislation*
- James C. Scott. (1972) *Comparative Political Corruption*
- Pietro Semeraro,(2008) Trading in influence and Lobbying in the Spanish Criminal Code (http://www.pravri.hr/hr/zavodi/zkzm/documents/decennium.pdf)
- Mark Wahlgren Summers. (1993) *The Era of Good Stealings*, corruption in American politics 1868–1877
- Darrell M. West (2000), *Checkbook Democracy. How Money Corrupts Political Campaigns,* Northeastern University Press, Boston (Mass.) ISBN 1-55553-440-6
- Alexandra Wrage (2007) Bribery and Extortion: Undermining Business, Governments and Security (http://www.amazon.com/dp/0275996492)
- Woodward, C. Vann, ed. *Responses of the Presidents to Charges of Misconduct* (1975), American presidents from Washington to Lyndon Johnson

External links

- On Tackling Corruption: A Non-Economist's View (http://stirringpottage.blogspot.com/2010/04/tackling-corruption.html) by Alum Bati
- Global Integrity Report (http://report.globalintegrity.org/) – local reporting and scorecards on anti-corruption performance in 90+ countries
- United Nations Convention against Corruption at Law-Ref.org (http://law-ref.org/CORRUPTION/index.html) – fully indexed and crosslinked with other documents
- PolicyPitch (http://www.policypitch.com) seeks to hold politicians accountable for their actions by providing more transparency. It allows people to track and comment on local legislation, contact politicians, and propose your own policies.
- Money and Politics – Political Finance & Public Ethics (http://moneyandpolitics.org/) – links to news articles, resources and handbooks on political corruption, political finance and campaign finance issues around the world

- UNODC – United Nations Office on Drugs and Crime – on corruption (http://www.unodc.org/unodc/en/corruption/index.html)
- UNODC corruption campaign – Your NO counts! (http://www.unodc.org/yournocounts/)
- World Bank anti-corruption page (http://www1.worldbank.org/publicsector/anticorrupt/)
- The World Bank's Private Sector Development Blog on Corruption (http://psdblog.worldbank.org/psdblog/corruption/index.html)
- World Bank's Worldwide Governance Indicators (http://www.govindicators.org/) Worldwide ratings of country performances on six governance dimensions from 1996 to present.
- Corruption Literature Review (http://web.worldbank.org/WBSITE/EXTERNAL/TOPICS/EXTPUBLICSECTORANDGOVERNANCE/EXTANTICORRUPTION/0,,contentMDK:20850389~pagePK:210058~piPK:210062~theSitePK:384455,00.html) World Bank Literature Review.
- UNICORN: A Global Trade Union Anti-corruption Network (http://www.againstcorruption.org/), based at Cardiff University
- The Committee for Fighting Corruption (Inter-Regional non-governmental organisation) (http://www.com-cor.ru/en)
- Corruption Perceptions Index 2009 Map (http://www.up2maps.net/maps/WorldIsMine/World/corruption_perceptions_index_2009.html) Source Transparency International
- SamuelGriffith.org.au (http://www.samuelgriffith.org.au/papers/html/volume8/v8chap7.htm), McGrath, Amy. Chapter Seven "One Vote, One Value: Electoral Fraud in Australia". Proceedings of the Eighth Conference of The Samuel Griffith Society.
- National Seminar on political corruption in India (http://www.scribd.com/doc/44708323/Two-Day-National-Seminar-Sponsored-by-UGC-on-Political-Corruption-in-India). 20–21 January 2011.
- Reducing corruption in public governance : Rhetoric to reality (http://www.scribd.com/doc/45735344)
- Prevention: An Effective Tool to Reduce Corruption (http://www.unodc.org/pdf/crime/gpacpublications/cicp2.pdf)
- Reducing corruption at the local level (http://www.ucl.ac.uk/dpu-projects/drivers_urb_change/urb_governance/pdf_trans_corrupt/WorldBank_Gonzalez_2000_Reducing_Corruption.pdf)
- Corrupt Cities : A Practical Guide to Cure and Prevention (162 pages (http://www.fpdl.ro/public/training_manuals/CORRUPT CITIES/CORRUPT CITIES EN/EN Corrupt Cities.pdf)
- Corruption through the programme cycle by TI (http://www.google.co.in/url?sa=t&source=web&cd=64&ved=0CDAQFjADODw&url=http://tistats.transparency.org/index.php/content/download/50526/808878/Hum_Hndbk_Corrpt_Progr_Cycl.pdf&rct=j&q=reducing corruption practical guides filetype:pdf&ei=T5IQTdHTO8rnrAean4zaCw&usg=AFQjCNGcDXt01HAHUIYNBDCF62Fpr7K6Mw&sig2=269zK17dGDZae2qohiqWkg)
- Understanding and preventing police corruption : Lessons from the literature (http://rds.homeoffice.gov.uk/rds/prgpdfs/fprs110.pdf)
- Resources on corruption, environment and development (http://www.environmentportal.in/category/thesaurus/governance-and-institutions/corruption)
- Index of Economic Freedom (http://www.heritage.org/index/download)

Ministry of Environment and Forests (India)

The **Ministry of Environment and Forests** (**MoEF**) is an Indian government ministry. The Minister of Environment and Forests holds cabinet rank as a member of the Council of Ministers. The ministry portfolio is currently held by Jayanthi Natarajan, Union Minister of Environment and Forests, who is assisted by two Ministers of State.

The ministry is responsible for planning, promoting, coordinating, and overseeing the implementation of environmental and forestry programmes in the country. The main activities undertaken by the ministry include conservation and survey of the flora of India and fauna of India, forests and other wilderness areas; prevention and control of pollution; afforestation, and land degradation mitigation. It is responsible for the administration of the national parks of India.

Organization

- **Subordinate offices**
  - Andaman & Nicobar Islands Forest and Plantation Development Corporation (Public Sector Undertaking)
  - Animal Welfare Board of India
  - Botanical Survey of India (BSI), Kolkata
  - Central Pollution Control Board
  - Central Zoo Authority
  - Directorate of Forest Education (DFE), Dehradun
  - Forest Survey of India (FSI), Dehradun
  - Indira Gandhi National Forest Academy (IGNFA), Dehradun
  - National Afforestation and Eco-Development Board
  - National Biodiversity Authority
  - National Ganga River Basin Authority
  - National Institute of Animal Welfare
  - National Museum of Natural History (NMNH), New Delhi
  - National Tiger Conservation Authority
  - National Zoological Park (NZP), New Delhi
  - Zoological Survey of India (ZSI), Kolkata
- **Autonomous institutions**
  - Central Pollution Control Board (CPCB), Delhi
  - Central Zoo Authority (CZA), New Delhi
  - G.B. Pant Institute of Himalayan Environment and Development, Almora
  - Indian Council of Forestry Research and Education (ICFRE), Dehradun
    - Arid Forestry Research Institute (AFRI), Jodhpur
    - Forest Research Institute (FRI), Dehradun
      - Centre for Forestry Research and Human Resource Development (CFRHRD), Chhindwara
    - Institute of Forest Genetics and Tree Breeding (IFGTB), Coimbatore
    - Institute (TFRI), Jabalpur
      - Centre for Social Forestry and Eco-habilitation (CSFE), Allahabad
    - Wildlife Institute of India WII

List of environment ministers of India

Period	Minister
2011-current	Jayanthi Natarajan
2009-2011	Jairam Ramesh

See also

- Awaaz Foundation - Conservation and environmental NGO
- A Pledge for Clean and Green India

External links

- Official website [1]
- Resources on Ministry of Environment and Forests (MoEF) [2]

It has some influence on agricultural ministry of India

References

[1] http://envfor.nic.in/
[2] http://www.indiaenvironmentportal.org.in/category/thesaurus/governance-and-institutions/governance/government-india-goi/ministry-environment-

Hill station

A **hill station** is a town located at a higher elevation than the nearby plain or valley. The term was used mostly in colonial Asia (particularly India), but also in Africa (albeit rarely), for towns founded by European colonial rulers as refuges from the summer heat, up where temperatures are cooler. In the Indian context most hill stations are at an altitude of approximately between 1,000 and 2,500 metres (3,500 to 7,500 feet); very few are outside this range.

Antsirabe

List of hill stations

This is a list of hill stations in various countries, which are mainly in Asia.

This list is incomplete.

Africa

Madagascar

- Antsirabe

Morocco

- Ifrane

Ifrane

Asia

Burma

- Pyin U Lwin
- Kalaw

Cambodia

- Bokor Hill Station

Saputara

Indonesia

- Bandung in West Java
- Garut in West Java
- Kaliurang in Central Java
- Sukabumi in West Java

Malaysia

- Bukit Tinggi
- Cameron Highlands
- Fraser's Hill
- Genting Highlands - founded following Malaysian independence.
- Maxwell Hill
- Penang Hill

Kalimpong

Pakistan (by province)

- Azad Kashmir
  - Muzaffarabad
- Khyber Pakhtunkhwa: Most of the hill stations of the KP are in the Galiyat region, which also extends into the Murree Tehsil of the Punjab province.
  - Abbottabad, Ayubia, Behrain, Chitral, Galyat, Kalam Valley, Malam Jabba, Nathia Gali, Patriata, Swat
- Punjab: All the hill stations listed here are in the Galiyat region. Although the Galiyat area is primarily in the Khyber Pakhtunkhwa, it also extends into the Murree Tehsil of the Punjab province. The largest hill station of the Galiyat is the town of Murree.

- Bhurban, Charra Pani, Murree, Patriata
- Northern Areas
  - Gilgit, Hunza, Skardu
- Sindh
  - Gorakh Hills

Philippines

- Baguio City

Sri Lanka

- Nuwara Eliya
- Ambewela

Vietnam

- Da Lat
- Sa Pa
- Tam Dao
- Ba Na

Garut

Cameron Highlands

Baguio City

Nuwara Eliya

Oceania

Australia

- Mount Macedon

References

- Barbara Crossette - *The Great Hill Stations of Asia*. ISBN 0-465-01488-7
- Kennedy, Dane.*The Magic Mountains: Hill Stations and the British Raj* (Full text, searchable) [1] Berkeley: University of California Press, c1996. ISBN 0520201884 ISBN 978-0520201880 Product Dimensions: 9.3 x 6.4 x 1.1 inches, Shipping Weight: 1.4 pounds [2]

Da Lat

Genting Highlands, a hill station founded after the independence of Malaysia.

References

[1] http://ark.cdlib.org/ark:/13030/ft396nb1sf/
[2] http://www.amazon.com/dp/0520201884

Indian independence movement

British Indian Empire

Colonial India	
Portuguese India	1510–1961
Dutch India	1605–1825
Danish India	1620–1869
French India	1759–1954
British India 1613–1947	
East India Company	1612–1757
Company rule in India	1757–1857
British Raj	1858–1947
British rule in Burma	1824–1942
Princely states	1765–1947
Partition of India	1947

The term **Indian independence movement** encompasses a wide spectrum of political organizations, philosophies, and movements which had the common aim of ending first British East India Company rule, and then British colonial authority, in parts of South Asia. The term is sometimes misleadingly used as the freedom movement for the country known as India after 1947. In fact, it also includes the Pakistan Movement as a result of minorities' and Muslims' struggle for equal rights as opposed to the Hindu-centric ideologies of Indian National Congress. The term incorporates various national and regional campaigns, agitations and efforts of both nonviolent and militant philosophy.

The first organized militant movements were in Bengal, but it later took political stage in the form of a mainstream movement in the then newly-formed Indian National Congress (INC), with prominent moderate leaders seeking only their basic right to appear for civil service examinations, as well as more rights, economic in nature, for the people of the soil. The early part of the 20th century saw a more radical approach towards political independence proposed by leaders such as the Lal Bal Pal and Sri Aurobindo. Militant nationalism also emerged in the first decades, culminating in the failed Indo-German Pact and Ghadar Conspiracy during the First World War.

The last stages of the freedom struggle from the 1920s onward saw Congress adopt Mohandas Karamchand Gandhi's policy of nonviolence and civil resistance, Muhammad Ali Jinnah's constitutional struggle for the rights of minorities in India, and several other campaigns. Some figures, such as Netaji Subhash Chandra Bose, later came to adopt a military approach to the movement, and others like Swami Sahajanand Saraswati who along with political freedom wanted economic freedom for India's peasants and toiling masses. Poets like Allama Iqbal used literature, poetry and speech as a tool for political awareness. The period of World War II saw the peak of movements such as the Quit India movement (led by Gandhi) and the Indian National Army (INA) movement (led by Bose).

These various movements led to the formation of the independent Dominions of India and Pakistan in 1947. India remained a Dominion of the Crown until 26 January 1950, when the *Constitution of India* came into force, establishing the Republic of India. Pakistan continued as a Dominion until 1956, when it too became a republic. In 1971, the Pakistani Civil War broke out, which led to war between Pakistan and India; the result of this conflict saw the splintering-off of East Pakistan as the new nation of Bangladesh. Several issues like Kashmir still remain disputed as an aftermath of the independence movement.

The Indian independence movement was a mass-based movement that encompassed various sections of society at the time. It also underwent a process of constant ideological evolution.[1] Although the basic ideology of the movement was anti-colonial, it was supported by a vision of independent capitalist economic development coupled with a secular, democratic, republican, and civil-libertarian political structure.[2] After the 1930s, the movement took on a strong socialist orientation, due to the increasing influence of left-wing elements in the INC as well as the rise and growth of the Communist Party of India.[1] On the other hand, due to INC's policies, Muslim League was formed to protect the rights of Muslims in Indian Sub-continent against INC and to present its voice to the British government.

Background (1757–1885)

Early British colonialism in India

European traders first reached Indian shores with the arrival of the Portuguese explorer Vasco da Gama in 1498 at the port of Calicut, in search of the lucrative spice trade. Just over a century later, the Dutch and English established trading outposts on the subcontinent, with the first English trading post set up at Surat in 1612.[3] Over the course of the seventeenth and early eighteenth centuries, the British[4] defeated militarily the Portuguese and Dutch, but remained in conflict with the French, who had by then sought to establish themselves in the subcontinent. The decline of the Mughal empire in the first half of the eighteenth century provided the British with a firm foothold in Indian politics.[5] After the Battle of Plassey in 1757, during which a British East India Company army under Robert Clive defeated Siraj-ud-Daula (the Nawab of Bengal), the Company established itself as a major player in Indian affairs, and soon after gained administrative rights over the regions of Bengal, Bihar, and Orissa following the Battle of Buxar in 1765.[6] After the defeat of Tipu Sultan, most of South India was now either under the Company's direct rule, or under its indirect political control as part of one of the princely states. The Company subsequently gained control of regions ruled by the Maratha Empire, after defeating them in a series of wars. Punjab was annexed in 1849 after the defeat of the Sikh armies in the First (1845–46) and Second (1848–49) Anglo-Sikh Wars.

Robert Clive, 1st Baron Clive with Mir Jafar after the Battle of Plassey

In 1835 English was made the medium of instruction in India's schools. Western-educated Hindu elites sought to rid Hinduism of controversial social practices, including the *varna* (caste) system, child marriage, and *sati*. Literary and debating societies initiated in Calcutta (Kolkata) and Bombay (Mumbai) became forums for open political discourse.

Even while these modernising trends influenced Indian society, Indians increasingly despised British rule. With the British now dominating most of the subcontinent, they grew increasingly abusive of local customs by, for example, staging parties in mosques, dancing to the music of regimental bands on the terrace of the Taj Mahal, using whips to force their way through crowded bazaars (as recounted by General Henry Blake), and mistreating Indians (including the sepoys). In the years after the annexation of Punjab in 1849, several mutinies broke out among the sepoys; these were put down by force.

The rebellion of 1857 and its consequences

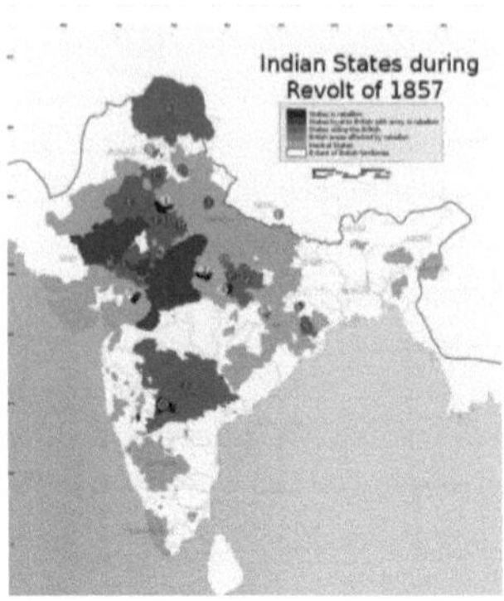

States during the rebellion

The Indian rebellion of 1857 was a period of uprising in northern and central India against the British East India Company's rule. The conditions of service in the Company's army and cantonments increasingly came into conflict with the religious beliefs and prejudices of the sepoys.Chandra 1989, p. 33 The predominance of members from the upper castes in the army, perceived loss of caste due to overseas travel, and rumours of secret designs of the Government to convert them to Christianity led to deep discontentment among the sepoys.Chandra 1989, p. 34 The sepoys were also disillusioned by their low salaries and the racial discrimination practised by British officers in matters of promotion and privileges.name="Chandra 34" The indifference of the British towards leading native Indian rulers such as the Mughals and ex-Peshwas and the annexation of Oudh were political factors triggering dissent amongst Indians. The Marquess of Dalhousie's policy of annexation, the doctrine of lapse (or escheat) applied by the British, and the projected removal of the descendants of the Great Mughal from their ancestral palace at Red Fort to the Qutb (near Delhi) also angered some people.

Secundra Bagh after the 93rd Highlanders and 4th Punjab regiment fought the rebels, November 1857

The final spark was provided by the rumoured use of cow and pig fat in the newly-introduced Pattern 1853 Enfield rifle cartridges. Soldiers had to bite the cartridges with their teeth before loading them into their rifles, and the reported presence of cow and pig fat was offensive to both Hindu and Muslim soldiers."The Uprising of 1857" [7]. Library of Congress. Retrieved 2009-11-10. On 10 May 1857, the sepoys at Meerut broke rank and turned on their commanding officers, killing some of them. They then reached Delhi on May 11, set the Company's toll house afire, and marched into the Red Fort, where they asked the Mughal emperor, Bahadur Shah II, to become their leader and reclaim his throne. The emperor was reluctant at first, but eventually agreed and was proclaimed *Shehenshah-e-Hindustan* by the rebels.Chandra 1989, p. 31 The rebels also murdered much of the European, Eurasian, and Christian population of the city.David, S (202) The India Mutiny, Penguin P122

Revolts broke out in other parts of Oudh and the North-Western Provinces as well, where civil rebellion followed the mutinies, leading to popular uprisings.Chandra 1989, p. 35 The British were initially caught off-guard and were thus slow to react, but eventually responded with force. The lack of effective organisation among the rebels, coupled with the military superiority of the British, brought a rapid end to the rebellion.Chandra 1989, pp. 38–39 The British fought the main army of the rebels near Delhi, and after prolonged fighting and a siege, defeated them and retook the city on 20 September 1857.Chandra 1989, p. 39 Subsequently, revolts in other centres were also crushed. The last significant battle was fought in Gwalior on 17 June 1858, during which Rani Lakshmi Bai was killed. Sporadic fighting and guerrilla warfare, led by Tantia Tope, continued until 1859, but most of the rebels were eventually subdued.

The Indian Rebellion of 1857 was a major turning point in the history of modern India. While affirming the military and political power of the British,Heehs 1998, p. 32 it led to significant change in how India was to be controlled by them. Under the Government of India Act 1858, the Company was deprived of its involvement in ruling India, with its territory being transferred to the direct authority of the British government. At the apex of the new system was a

Cabinet minister, the Secretary of State for India, who was to be formally advised by a statutory council;Heehs 1998, pp. 47–48 the Governor-General of India (Viceroy) was made responsible to him, while he in turn was responsible to the British Parliament for British rule. In a royal proclamation made to the people of India, Queen Victoria promised equal opportunity of public service under British law, and also pledged to respect the rights of the native princes.Heehs 1998, p. 48 The British stopped the policy of seizing land from the princes, decreed religious tolerance, and began to admit Indians into the civil service (albeit mainly as subordinates). However, they also increased the number of British soldiers in relation to native Indian ones, and only allowed British soldiers to handle artillery. Bahadur Shah was exiled to Rangoon (Yangon), Burma (Myanmar), where he died in 1862.

In 1876, Queen Victoria took the additional title of Empress of India.

Rise of organized movements

The decades following the Rebellion were a period of growing political awareness, manifestation of Indian public opinion, and emergence of Indian leadership at both national and provincial levels. Dadabhai Naoroji formed the East India Association in 1867, and Surendranath Banerjee founded the Indian National Association in 1876.

Inspired by a suggestion made by A. Hume, a retired British civil servant, seventy-three Indian delegates met in Bombay in 1885 and founded the Indian National Congress. They were mostly members of the upwardly mobile and successful western-educated provincial elites, engaged in professions such as law, teaching, and journalism. At its inception, the Congress had no well-defined ideology and commanded few of the resources essential to a political organization. Instead, it functioned more as a debating society that met annually to express its loyalty to the British Raj, and passed numerous resolutions on less controversial issues such as civil rights or opportunities in government (especially in the civil service). These resolutions were submitted to the Viceroy's government and occasionally to the British Parliament, but the Congress's early gains were meagre. Despite its claim to represent all India, the Congress voiced the interests of urban elites; the number of participants from other social and economic backgrounds remained negligible.

The influence of socio-religious groups such as *Arya Samaj* (started by Swami Dayanand Saraswati) and *Brahmo Samaj* (founded by Raja Ram Mohan Roy and others) became evident in pioneering reforms of Indian society. The work of men like Swami Vivekananda, Ramakrishna Paramhansa, Sri Aurobindo, Subramanya Bharathy, Bankim Chandra Chatterjee, Sir Syed Ahmed Khan, Rabindranath Tagore, and Dadabhai Naoroji, as well as women such as the Scots–Irish Sister Nivedita, spread the passion for rejuvenation and freedom. The rediscovery of India's indigenous history by several European and Indian scholars also fed into the rise of nationalism among Indians.

Rise of Indian nationalism (1885–1905)

By 1900, although the Congress had emerged as an all-India political organization, its achievement was undermined by its singular failure to attract Muslims, who felt that their representation in government service was inadequate. Attacks by Hindu reformers against religious conversion, cow slaughter, and the preservation of Urdu in Arabic script deepened their concerns of minority status and denial of rights if the Congress alone were to represent the people of India. Sir Syed Ahmed Khan launched a movement for Muslim regeneration that culminated in the founding in 1875 of the Muhammadan Anglo-Oriental College at Aligarh, Uttar Pradesh (renamed Aligarh Muslim University in 1920). Its objective was to educate wealthy students by emphasizing the compatibility of Islam with modern western knowledge. The diversity among India's Muslims, however, made it impossible to bring about uniform cultural and intellectual regeneration.

The nationalistic sentiments among Congress members led to the movement to be represented in the bodies of government, to have a say in the legislation and administration of India. Congressmen saw themselves as loyalists, but wanted an active role in governing their own country, albeit as part of the Empire. This trend was personified by Dadabhai Naoroji, who went as far as contesting, successfully, an election to the British House of Commons, becoming its first Indian member.

Bal Gangadhar Tilak

Bal Gangadhar Tilak was the first Indian nationalist to embrace *Swaraj* as the destiny of the nation. Tilak deeply opposed the then British education system that ignored and defamed India's culture, history and values. He resented the denial of freedom of expression for nationalists, and the lack of any voice or role for ordinary Indians in the affairs of their nation. For these reasons, he considered Swaraj as the natural and only solution. His popular sentence "Swaraj is my birthright, and I shall have it" became the source of inspiration for Indians.

In 1907, the Congress was split into two factions. The *radicals* led by Tilak advocated civil agitation and direct revolution to overthrow the British Empire and the abandonment of all things British. The *moderates* led by leaders like Dadabhai Naoroji and Gopal Krishna Gokhale on the other hand wanted reform within the framework of British rule. Tilak was backed by rising public leaders like Bipin Chandra Pal and Lala Lajpat Rai, who held the same point of view. Under them, India's three great states - Maharashtra, Bengal and Punjab shaped the demand of the people and India's nationalism. Gokhale criticized Tilak for encouraging acts of violence and disorder. But the Congress of 1906 did not have public membership, and thus Tilak and his supporters were forced to leave the party.

But with Tilak's arrest, all hopes for an Indian offensive were stalled. The Congress lost credit with the people. A Muslim deputation met with the Viceroy, Minto (1905–10), seeking concessions from the impending constitutional reforms, including special considerations in government service and electorates. The British recognized some of the Muslim League's petitions by increasing the number of elective offices reserved for Muslims in the Indian Councils Act 1909. The Muslim League insisted on its separateness from the Hindu-dominated Congress, as the voice of a "nation within a nation."

Partition of Bengal, 1905

In July 1905, Lord Curzon, the Viceroy and Governor-General (1899–1905), ordered the partition of the province of Bengal supposedly for improvements in administrative efficiency in the huge and populous region. It also had justifications due to increasing conflicts between Muslims and dominant Hindu regimes in Bengal. However the Indians viewed the partition as an attempt by the British to disrupt the growing national movement in Bengal and divide the Hindus and Muslims of the region. The Bengali Hindu intelligentsia exerted considerable influence on local and national politics. The partition outraged Bengalis. Not only had the government failed to consult Indian public opinion, but the action appeared to reflect the British resolve to divide and rule. Widespread agitation ensued in the streets and in the press, and the Congress advocated boycotting British products under the banner of *swadeshi*. Hindus showed unity by tying Rakhi on each other's wrists and observing *Arandhan* (not cooking any food). During this time Bengali Hindu nationalists begin writing virulent newspaper articles and were charged with sedition. Brahmabhandav Upadhyay, a Hindu newspaper editor who helped Tagore establish his school at Shantiniketan, was imprisoned and the first martyr to die in British custody in the 20th century struggle for independence.

During the partition of Bengal new methods of struggle were adopted. These led to *swadeshi* and boycott movements. The Congress-led boycott of British goods was so successful that it unleashed anti-British forces to an extent unknown since the Sepoy Rebellion. A cycle of violence and repression ensued in some parts of the country (see Alipore bomb case). The British tried to mitigate the situation by announcing a series of constitutional reforms in 1909 and by appointing a few moderates to the imperial and provincial councils. In what the British saw as an additional goodwill gesture, in 1911 King-Emperor George V visited India for a *durbar* (a traditional court held for subjects to express fealty to their ruler), during which he announced the reversal of the partition of Bengal and the transfer of the capital from Calcutta to a newly planned city to be built immediately south of Delhi, which later became New Delhi. However, the ceremony of transfer on 23 December 1912 was marked by the attempt to assassinate the then Viceroy, Lord Hardinge, in what came to be known as the Delhi-Lahore conspiracy.

All India Muslim League

The All India Muslim League was founded by the All India Muhammadan Educational Conference at Dhaka (now Bangladesh), in 1906, in the context of the circumstances that were generated over the partition of Bengal in 1905. Being a political party to secure the interests of the Muslim diaspora in British India, the Muslim League played a decisive role during the 1940s in the Indian independence movement and developed into the driving force behind the creation of Pakistan in the Indian subcontinent.[8]

Muhammad Ali Jinnah's struggle for Hindu-Muslim Unity and "Home Rule"

In 1906, Muhammad Ali Jinnah joined the Indian National Congress, which was the largest Indian political organization. Like most of the Congress at the time, Jinnah did not favour outright independence, considering British influences on education, law, culture and industry as beneficial to India. Jinnah became a member on the sixty-member Imperial Legislative Council. The council had no real power or authority, and included a large number of un-elected pro-Raj loyalists and Europeans. Nevertheless, Jinnah was instrumental in the passing of the *Child Marriages Restraint Act*, the legitimization of the Muslim waqf (religious endowments) and was appointed to the Sandhurst committee, which helped establish the Indian Military Academy at Dehra Dun.[9] During World War I, Jinnah joined other Indian moderates in supporting the British war effort, hoping that Indians would be rewarded with political freedoms.

Muhammad Ali Jinnah, as a young lawyer.

Jinnah had initially avoided joining the All India Muslim League, founded in 1906, regarding it as too Muslim oriented. However he decided to provide leadership to the Muslim minority. Eventually, he joined the league in 1913 and became the president at the 1916 session in Lucknow. Jinnah was the architect of the 1916 Lucknow Pact between the Congress and the League, bringing them together on most issues regarding self-government and presenting a united front to the British. Jinnah also played an important role in the founding of the All India Home Rule League in 1916. Along with political leaders Annie Besant and Tilak, Jinnah demanded "home rule" for India—the status of a self-governing dominion in the Empire similar to Canada, New Zealand and Australia. He headed the League's Bombay Presidency chapter.

In 1924 Jinnah reorganized the Muslim League, of which he had been president since 1916, and devoted the next seven years attempting to bring about unity among the disparate ranks of Muslims and to develop a rational formula to effect a Hindu-Muslim settlement, which he considered the pre condition for Indian freedom. He attended several unity conferences, wrote the Delhi Muslim Proposals in 1927, pleaded for the incorporation of the basic Muslim demands in the Nehru report, and formulated the Fourteen Points.[10]

World War I

World War I began with an unprecedented outpouring of loyalty and goodwill towards the United Kingdom from within the mainstream political leadership, contrary to initial British fears of an Indian revolt. India contributed massively to the British war effort by providing men and resources. About 1.3 million Indian soldiers and labourers served in Europe, Africa, and the Middle East, while both the Indian government and the princes sent large supplies of food, money, and ammunition. However, Bengal and Punjab remained hotbeds of anti colonial activities. Nationalism in Bengal, increasingly closely linked with the unrests in Punjab, was significant enough to nearly paralyse the regional administration.[11] [12] Also from the beginning of the war, expatriate Indian population, notably from United States, Canada, and Germany, headed by the Berlin Committee and the Ghadar Party, attempted to trigger insurrections in India on the lines of the 1857 uprising with Irish Republican, German and Turkish help in a massive conspiracy that has since come to be called the Hindu–German Conspiracy[13] [14] [15] This conspiracy also attempted to rally Afghanistan against British India.[16] A number of failed attempts were made at mutiny, of which the February mutiny plan and the Singapore mutiny remains most notable. This movement was suppressed by means of a massive international counter-intelligence operation and draconian political acts (including the Defence of India act 1915) that lasted nearly ten years.[17] [18]

In the aftermath of the World War I, high casualty rates, soaring inflation compounded by heavy taxation, a widespread influenza epidemic, and the disruption of trade during the war escalated human suffering in India. The Indian soldiers smuggled arms into India to overthrow the British rule. The pre-war nationalist movement revived as moderate and extremist groups within the Congress submerged their differences in order to stand as a unified front. In 1916, the Congress succeeded in forging the Lucknow Pact, a temporary alliance with the Muslim League over the issues of devolution of political power and the future of Islam in the region.

The British themselves adopted a "carrot and stick" approach in recognition of India's support during the war and in response to renewed nationalist demands. In August 1917, Edwin Montagu, the secretary of state for India, made the historic announcement in Parliament that the British policy for India was "increasing association of Indians in every branch of the administration and the gradual development of self-governing institutions with a view to the progressive realization of responsible government in India as an integral part of the British Empire." The means of achieving the proposed measure were later enshrined in the Government of India Act 1919, which introduced the principle of a dual mode of administration, or diarchy, in which both elected Indian legislators and appointed British officials shared power. The act also expanded the central and provincial legislatures and widened the franchise considerably. Diarchy set in motion certain real changes at the provincial level: a number of non-controversial or "transferred" portfolios, such as agriculture, local government, health, education, and public works, were handed over to Indians, while more sensitive matters such as finance, taxation, and maintaining law and order were retained by the provincial British administrators.

Gandhi arrives in India

Mohandas Karamchand Gandhi (also known as Mahatma Gandhi), had been a prominent leader of the anti-Apartheid movement in South Africa, and had been a vocal opponent of basic discrimination and abusive labour treatment as well as suppressive police control such as the Rowlatt Acts. During these protests, Gandhi had perfected the concept of *satyagraha*, which had been inspired by the philosophy of Baba Ram Singh (famous for leading the Kuka Movement in the Punjab in 1872). The end of the protests in South Africa saw oppressive legislation repealed and the release of political prisoners by General Jan Smuts, head of the South African Government of the time.

Gandhi returned to India, on 6 January 1915 and initially entered the political fray not with calls for a nation-state, but in support of the unified commerce-oriented territory that the Congress Party had been asking for. Gandhi believed that the industrial development and educational development that the Europeans had brought with them were required to alleviate many of India's problems. Gopal Krishna Gokhale, a veteran Congressman and Indian leader, became Gandhi's mentor. Gandhi's ideas and strategies of non-violent civil disobedience initially appeared impractical to some Indians and Congressmen. In Gandhi's own words, "civil disobedience is civil breach of unmoral statutory enactments." It had to be carried out non-violently by withdrawing cooperation with the corrupt state. Gandhi's ability to inspire millions of common people became clear when he used satyagraha during the anti-Rowlatt Act protests in Punjab. Gandhi had great respect to Lokmanya Tilak. His programmes were all inspired by Tilak's "Chatusutri" programme.

Gandhi's vision would soon bring millions of regular Indians into the movement, transforming it from an elitist struggle to a national one. The nationalist cause was expanded to include the interests and industries that formed the economy of common Indians. For example, in Champaran, Bihar, the Congress Party championed the plight of desperately poor sharecroppers and landless farmers who were being forced to pay oppressive taxes and grow cash crops at the expense of the subsistence crops which formed their food supply. The profits from the crops they grew were insufficient to provide for their sustenance.

The positive impact of reform was seriously undermined in 1919 by the Rowlatt Act, named after the recommendations made the previous year to the Imperial Legislative Council by the Rowlatt Commission, which had been appointed to investigate what was termed the "seditious conspiracy" and the German and Bolshevik involvement in the militant movements in India.[19] [20] [21] The Rowlatt Act, also known as the Black Act, vested the Viceroy's government with extraordinary powers to quell sedition by silencing the press, detaining the political activists without trial, and arresting any individuals suspected of sedition or treason without a warrant. In protest, a nationwide cessation of work (*hartal*) was called, marking the beginning of widespread, although not nationwide, popular discontent. The agitation unleashed by the acts culminated on 13 April 1919, in the Jallianwala Bagh massacre (also known as the Amritsar Massacre) in Amritsar, Punjab. The British military commander, Brigadier-General Reginald Dyer, blocked the main entrance-cum-exit, and ordered his soldiers to fire into an unarmed and unsuspecting crowd of some 5,000 men, women and children. They had assembled at Jallianwala Bagh, a walled courtyard in defiance of the ban. A total of 1,651 rounds were fired, killing 379 people (as according to an official British commission; Indian estimates ranged as high as 1,499[22]) and wounding 1,137 in the episode, which dispelled wartime hopes of home rule and goodwill in a frenzy of post-war reaction.

The non-cooperation movements

It can be argued that the independence movement, even towards the end of First World War, was far removed from the masses of India, focusing essentially on a unified commerce-oriented territory and hardly a call for a united nation. That came in the 1930s with the entry of Mohandas Karamchand Gandhi into Indian Politics in 1915.

The first non-cooperation movement

At the Calcutta session of the Congress in September 1920, Gandhi convinced other leaders of the need to start a non-cooperation movement in support of Khilafat as well as for swaraj (self rule). The first satyagraha movement urged the use of khadi and Indian material as alternatives to those shipped from Britain. It also urged people to boycott British educational institutions and law courts; resign from government employment; refuse to pay taxes; and forsake British titles and honours. Although this came too late to influence the framing of the new *Government of India Act 1919*, the movement enjoyed widespread popular support, and the resulting unparalleled magnitude of disorder presented a serious challenge to foreign rule. However, Gandhi called off the movement following the Chauri Chaura incident, which saw the death of twenty-two policemen at the hands of an angry mob.

Membership in the party was opened to anyone prepared to pay a token fee, and a hierarchy of committees was established and made responsible for discipline and control over a hitherto amorphous and diffuse movement. The party was transformed from an elite organization to one of mass national appeal and participation.

Gandhi was sentenced in 1922 to six years of prison, but was released after serving two. On his release from prison, he set up the Sabarmati Ashram in Ahmedabad, on the banks of river Sabarmati, established the newspaper *Young India*, and inaugurated a series of reforms aimed at the socially disadvantaged within Hindu society — the rural poor, and the untouchables.

This era saw the emergence of new generation of Indians from within the Congress Party, including C. Rajagopalachari, Jawaharlal Nehru, Vallabhbhai Patel, Subhash Chandra Bose and others- who would later on come to form the prominent voices of the Indian independence movement, whether keeping with Gandhian Values, or diverging from it.

The Indian political spectrum was further broadened in the mid-1920s by the emergence of both moderate and militant parties, such as the Swaraj Party, Hindu Mahasabha, Communist Party of India and the Rashtriya Swayamsevak Sangh. Regional political organizations also continued to represent the interests of non-Brahmins in Madras, Mahars in Maharashtra, and Sikhs in Punjab. However, people like Mahakavi Subramanya Bharathi, Vanchinathan and Neelakanda Brahmachari played a major role from Tamil Nadu in both freedom struggle and fighting for equality for all castes and communities.

Purna Swaraj

Following the rejection of the recommendations of the Simon Commission by Indians, an all-party conference was held at Bombay in May 1928. This was meant to instill a sense of resistance among people. The conference appointed a drafting committee under Motilal Nehru to draw up a constitution for India. The Calcutta session of the Indian National Congress asked the British government to accord dominion status to India by December 1929, or a countrywide civil disobedience movement would be launched. By 1929, however, in the midst of rising political discontent and increasingly violent regional movements, the call for complete independence from Britain began to find increasing grounds within the Congress leadership. Under the presidency of Jawaharlal Nehru at its historic Lahore session in December 1929, The Indian National Congress adopted a resolution calling for complete independence from the British. It authorised the Working Committee to launch a civil disobedience movement throughout the country. It was decided that 26 January 1930 should be observed all over India as the *Purna Swaraj* (total independence) Day. Many Indian political parties and Indian revolutionaries of a wide spectrum united to observe the day with honour and pride.

Karachi congress session-1931 A special session was held to endorse the Gandhi-Irwin or Delhi Pact. The goal of Purna swaraj was reiterated. Two resolutions were adopted-one on Fundamental rights and other on National Economic programme.which made the session perticularly memmorable.

This was the first time the congress spelt out what swaraj would mean for the masses.

Salt March and civil disobedience

Gandhi emerged from his long seclusion by undertaking his most famous campaign, a march of about 400 kilometers [240 miles] from his commune in Ahmadabad to Dandi, on the coast of Gujarat between 11 March and 6 April 1930. The march is usually known as the *Dandi March* or the *Salt Satyagraha*. At Dandi, in protest against British taxes on salt, he and thousands of followers broke the law by making their own salt from seawater. It took 24 days for him to complete this march. Every day he covered 10 miles and gave many speeches.

In April 1930 there were violent police-crowd clashes in Calcutta. Approximately 100,000 people were imprisoned in the course of the Civil disobedience movement (1930–31), while in Peshawar unarmed demonstrators were fired upon in the Qissa Khwani bazaar massacre. The latter event catapulted the then newly formed Khudai Khidmatgar movement (founder Khan Abdul Ghaffar Khan, the *Frontier Gandhi*) onto the National scene. While Gandhi was in jail, the first Round Table Conference was held in London in November 1930, without representation from the Indian National Congress. The ban upon the Congress was removed because of economic hardships caused by the satyagraha. Gandhi, along with other members of the Congress Working Committee, was released from prison in January 1931.

In March 1931, the Gandhi-Irwin Pact was signed, and the government agreed to set all political prisoners free (Although, some of the key revolutionaries were not set free and the death sentence for Bhagat Singh and his two comrades was not taken back which further intensified the agitation against Congress not only outside it but with in the Congress itself). In return, Gandhi agreed to discontinue the civil disobedience movement and participate as the sole representative of the Congress in the second Round Table Conference, which was held in London in September 1931. However, the conference ended in failure in December 1931. Gandhi returned to India and decided to resume the civil disobedience movement in January 1932.

For the next few years, the Congress and the government were locked in conflict and negotiations until what became the Government of India Act 1935 could be hammered out. By then, the rift between the Congress and the Muslim League had become unbridgeable as each pointed the finger at the other acrimoniously. The Muslim League disputed the claim of the Congress to represent all people of India, while the Congress disputed the Muslim League's claim to voice the aspirations of all Muslims.

Elections and the Lahore resolution

The Government of India Act 1935, the voluminous and final constitutional effort at governing British India, articulated three major goals: establishing a loose federal structure, achieving provincial autonomy, and safeguarding minority interests through separate electorates. The federal provisions, intended to unite princely states and British India at the centre, were not implemented because of ambiguities in safeguarding the existing privileges of princes. In February 1937, however, provincial autonomy became a reality when elections were held; the Congress emerged as the dominant party with a clear majority in five provinces and held an upper hand in two, while the Muslim League performed poorly.

Jinnah bhai with Gandhi, 1944.

In 1939, the Viceroy Linlithgow declared India's entrance into World War II without consulting provincial governments. In protest, the Congress asked all of its elected representatives to resign from the government. Jinnah, the president of the Muslim League, persuaded participants at the annual Muslim League session at Lahore in 1940 to adopt what later came to be known as the Lahore Resolution, demanding the division of India into two separate sovereign states, one Muslim, the other Hindu; sometimes referred to as Two Nation Theory. Although the idea of Pakistan had been introduced as early as 1930, very few had responded to it. However, the volatile political climate and hostilities between the Hindus and Muslims transformed the idea of Pakistan into a stronger demand.

Revolutionary activities

Bagha Jatin

Apart from a few stray incidents, the armed rebellion against the British rulers was not organized before the beginning of the 20th century. The Indian revolutionary underground began gathering momentum through the first decade of 1900s, with groups arising in Bengal, Maharastra, Orissa, Bihar, Uttar Pradesh, Punjab, and the then Madras Presidency including what is now called South India. More groups were scattered around India. Particularly notable movements arose in Bengal, especially around the Partition of Bengal in 1905, and in Punjab.[23] In the former case, it was the educated, intelligent and dedicated youth of the urban Middle Class *Bhadralok* community that came to form the "Classic" Indian revolutionary,[23] while the latter had an immense support base in the rural and Military society of the Punjab. Organisations like Jugantar and Anushilan Samiti had emerged in the 1900s. The revolutionary philosophies and movement made their presence felt during the 1905 Partition of Bengal. Arguably, the initial steps to organize the revolutionaries were taken by Aurobindo Ghosh, his brother Barin Ghosh, Bhupendranath Datta etc. when they formed the Jugantar party in April 1906.[24] Jugantar was created as an inner circle of the Anushilan Samiti which was already present in Bengal mainly as a revolutionary society in the guise of a fitness club.

The Anushilan Samiti and Jugantar opened several branches throughout Bengal and other parts of India and recruited young men and women to participate in the revolutionary activities. Several murders and looting were done, with many revolutionaries being captured and imprisoned. The Jugantar party leaders like Barin Ghosh and Bagha Jatin initiated making of explosives. Amongst a number of notable events of political terrorism were the Alipore bomb case, the Muzaffarpur killing tried several activists and many were sentenced to deportation for life, while Khudiram Bose was hanged. The founding of the India House and The Indian Sociologist under Shyamji Krishna Varma in London in 1905 took the radical movement to Britain itself. On 1 July 1909, Madan Lal Dhingra, an Indian student closely identified with India House in London shot dead William Hutt Curzon Wylie, a British M.P. in London. 1912 saw the Delhi-Lahore Conspiracy planned under Rash Behari Bose, an erstwhile Jugantar member, to assassinate the then Viceroy of India Charles Hardinge. The conspiracy culminated in an attempt to Bomb the Viceregal procession on 23 December 1912, on the occasion of transferring the Imperial Capital from Calcutta to Delhi. In the aftermath of this event, concentrated police and intelligence efforts were made by the British Indian police to destroy the Bengali and Punabi revolutionary underground, which came under intense pressure for sometime. Rash Behari successfully evaded capture for nearly three years. However, by the time that World War I opened in Europe, the revolutionary movement in Bengal (and Punjab) had revived and was strong enough to nearly paralyse the local administration.[11] [12] in 1914, Indian revolutionaries made conspiracies against British rule but the plan was failed and many revolutionaries sacrificed their life and others were arrested and sent to the Cellular Jail (Kalapani) in Andaman and Nicobar Islands. During the First World War, the revolutionaries planned to import arms and ammunitions from Germany and stage an armed revolution against the British.[25]

The Ghadar Party operated from abroad and cooperated with the revolutionaries in India. This party was instrumental in helping revolutionaries inside India catch hold of foreign arms.

After the First World War, the revolutionary activities began to slowly wane as it suffered major setbacks due to the arrest of prominent leaders. In the 1920s, some revolutionary activists began to reorganize.

Kakori Conspiracy (9 Aug 1925)

Ram Prasad Bismil's Photo

In order to overturn the British Rule through arms, the idea of the robbery was conceived by Ram Prasad Bismil and Ashfaqullah Khan who belonged to the Hindustan Republican Association (HRA, which became HSRA or Hindustan Socialist Republican Association in 1928) that was created to carry out revolutionary activities against the British Empire in India. The objective of the HRA was to conduct an armed revolution against the British government. The organization needed money for the supply of weaponry, and thus Bismil decided to loot a train on one of the Northern Railway lines.[26] The robbery plan was executed by Ram Prasad Bismil, Ashfaqulla Khan, Rajendra Lahiri, Chandrasekhar Azad, Sachindra Bakshi, Keshab Chakravarthy (fake name of K.B. Hedgewar), Manmathnath Gupta, Murari Sharma (fake name of Murari Lal Gupta), Mukundi Lal (Mukundi Lal Gupta),.[27] [28] [29] In this historical event 40 persons belonging to HRA were arrested and a Conspiracy case was filed in which 4 were sentenced to death and 16 others were given imprisonment varying from 2 years to life importation.

Hindustan Socialist Republican Association was formed under the leadership of Chandrasekhar Azad. Bhagat Singh and Batukeshwar Dutt threw a bomb inside the Central Legislative Assembly on 8 April 1929 protesting against the passage of the Public Safety Bill and the Trade Disputes Bill. Following the trial (Central Assembly Bomb Case), Bhagat Singh, Sukhdev and Rajguru were hanged in 1931. Allama Mashriqi founded Khaksar Tehreek in order to direct particularly the Muslims towards the independence movement.[30]

Surya Sen, along with other activists, raided the Chittagong armoury on 18 April 1930 to capture arms and ammunition and to destroy government communication system to establish a local governance. Pritilata Waddedar led an attack on a European club in Chittagong in 1932, while Bina Das attempted to assassinate Stanley Jackson, the Governor of Bengal inside the convocation hall of Calcutta University. Following the Chittagong armoury raid case, Surya Sen was hanged and several others were deported for life to the Cellular Jail in Andaman. The Bengal Volunteers started operating in 1928. On 8 December 1930, the Benoy-Badal-Dinesh trio of the party entered the secretariat Writers' Building in Kolkata and murdered Col. N. S. Simpson, the Inspector General of Prisons.

On 13 March 1940, Udham Singh shot Michael O'Dwyer, generally held responsible for the Amritsar Massacre, in London. However, as the political scenario changed in the late 1930s — with the mainstream leaders considering several options offered by the British and with religious politics coming into play — revolutionary activities gradually declined. Many past revolutionaries joined mainstream politics by joining Congress and other parties, especially communist ones, while many of the activists were kept under hold in different jails across the country.

The climax: WW2, Quit India, INA, INA trials and post-war revolts

Indians throughout the country were divided over World War II, as Linlithgow, without consulting the Indian representatives had unilaterally declared India a belligerent on the side of the allies. In opposition to Linlithgow's action, the entire Congress leadership resigned from the local government councils. However, many wanted to support the British war effort, and indeed the British Indian Army was one of the largest volunteer forces, numbering 205,000 men during the war.[31] Especially during the Battle of Britain, Gandhi resisted calls for massive civil disobedience movements that came from within as well as outside his party, stating he did not seek India's freedom out of the ashes of a destroyed Britain. However, like the changing fortunes of the war itself, the movement for freedom saw the rise of three movements that formed the climax of the 100-year struggle for independence.

The first of these, the Kakori conspiracy (9 August 1925) was done by the Indian youth under the leadership of Pandit Ram Prasad Bismil, second was the Azad Hind movement led by Netaji Subhash Chandra Bose, saw its inception early in the war and sought help from the Axis Powers. And the third one after 17 years of the first from the same date (9) saw its inception in August 1942 which was led by Lal Bahadur Shastri[32] and the common man resulting the failure of the Cripps' mission to reach a consensus with the Indian political leadership over the transfer of power after the war.

Quit India (9 Aug 1942)

The Quit India Movement *(Bharat Chhodo Andolan)* or the *August Movement* was a civil disobedience movement in India launched on 9 August 1942 in response to Gandhi's call for immediate independence of India and against sending Indians to World War II. He asked all the teachers to leave their school, and other Indians to leave away their respective jobs and take part in this movement. Due to Gandhi's political influence, request was followed on a massive proportion of the population.

At the outbreak of war, the Congress Party had during the Wardha meeting of the working-committee in September 1939, passed a resolution conditionally supporting the fight against fascism,[33] but were rebuffed when they asked for independence in return. In March 1942, faced with an increasingly dissatisfied sub-continent only reluctantly participating in the war, and deteriorations in the war situation in Europe and South East Asia, and with growing dissatisfactions among Indian troops- especially in Europe- and among the civilian population in the sub-continent, the British government sent a delegation to India under Stafford Cripps, in what came to be known as the Cripps' Mission. The purpose of the mission was to negotiate with the Indian National Congress a deal to obtain total co-operation during the war, in return of progressive devolution and distribution of power from the crown and the Viceroy to elected Indian legislature. However, the talks failed, having failed to address the key demand of a timeframe towards self-government, and of definition of the powers to be relinquished, essentially portraying an offer of limited dominion-status that was wholly unacceptable to the Indian movement.[34] To force the Raj to meet its demands and to obtain definitive word on total independence, the Congress took the decision to launch the Quit India Movement.

The aim of the movement was to bring the British Government to the negotiating table by holding the Allied War Effort hostage. The call for determined but passive resistance that signified the certitude that Gandhi foresaw for the movement is best described by his call to *Do or Die*, issued on 8 August at the Gowalia Tank Maidan in Bombay, since re-named *August Kranti Maidan* (August Revolution Ground). However, almost the entire Congress leadership, and not merely at the national level, was put into confinement less than twenty-four hours after Gandhi's speech, and the greater number of the Congress khiland were to spend the rest of the war in jail.

On 8 August 1942, the Quit India resolution was passed at the Bombay session of the All India Congress Committee (AICC). The draft proposed that if the British did not accede to the demands, a massive Civil Disobedience would be launched. However, it was an extremely controversial decision. At Gowalia Tank, Mumbai, Gandhi urged Indians to follow a non-violent civil disobedience. Gandhi told the masses to act as an independent nation and not to follow the orders of the British. The British, already alarmed by the advance of the Japanese army to the India–Burma border, responded the next day by imprisoning Gandhi at the Aga Khan Palace in Pune. The Congress Party's Working Committee, or national leadership was arrested all together and imprisoned at the Ahmednagar Fort. They also banned the party altogether. Large-scale protests and demonstrations were held all over the country. Workers remained absent en masse and strikes were called. The movement also saw widespread acts of sabotage, Indian under-ground organisation carried out bomb attacks on allied supply convoys, government buildings were set on fire, electricity lines were disconnected and transport and communication lines were severed. The Congress had lesser success in rallying other political forces, including the Muslim League under a single mast and movement. It did however, obtain passive support from a substantial Muslim population at the peak of the movement.The movement soon became a leaderless act of defiance, with a number of acts that deviated from Gandhi's principle of non-violence. In large parts of the country, the local underground organisations took over the movement. However, by 1943, *Quit India* had petered out.

The Indian National Army

The arbitrary entry of India into the war was strongly opposed by Netaji Subhash Chandra Bose, who had been elected President of the Congress twice, in 1938 and 1939. After lobbying against participation in the war, he resigned from Congress in 1939 and started a new party, the All India Forward Bloc. When war broke out, the Raj had put him under house arrest in Calcutta in 1940. However, he escaped and made his way through Afghanistan to Germany to seek Axis help to raise an army to fight the Raj. Here, he raised with Rommel's Indian POWs what came to be known as the Free India Legion. Bose made his way ultimately to Japanese South Asia where he formed what came to be known as the Azad Hind Government, a Provisional Free Indian Government in exile, and organized the Indian National Army with Indian POWs and Indian expatriates in South-East Asia, with the help of the Japanese. Its aim was to reach India as a fighting force that would build on public resentment to inspire revolts among Indian soldiers to defeat the Raj.

Netaji Subhash Chandra Bose.

The INA was to see action against the allies, including the British Indian Army, in the forests of Arakan, Burma and in Assam, laying siege on Imphal and Kohima with the Japanese 15th Army. During the war, the Andaman and Nicobar islands were captured by the Japanese and handed over by them to the INA. Bose renamed them *Shahid* (Martyr) and *Swaraj* (Independence).

The INA would ultimately fail, owing to disrupted logistics, poor arms and supplies from the Japanese, and lack of support and training. [35] The supposed death of Bose is seen as culmination of the entire Azad Hind Movement. Following the surrender of Japan, the troops of the INA were brought to India and a number of them charged with treason. However, Bose's actions had captured the public imagination and also turned the inclination of the native soldiers of the British Indian Forces from one of loyalty to the crown to support for the soldiers that the Raj deemed as collaborators.[36] [37]

After the war, the stories of the Azad Hind movement and its army that came into public limelight during the trials of soldiers of the INA in 1945 were seen as so inflammatory that, fearing mass revolts and uprisings — not just in India, but across its empire — the British Government forbade the BBC from broadcasting their story.[38] Newspapers reported the summary execution of INA soldiers held at Red Fort.[39] During and after the trial, mutinies broke out in the British Indian Armed forces, most notably in the Royal Indian Navy which found public support throughout India, from Karachi to Mumbai and from Vizag to Kolkata.[40] [41] [42] Many historians have argued that the INA, and the mutinies it inspired, were strong driving forces behind the transfer of power in 1947.[43] [44] [45]

Christmas Island Mutiny

After two Japanese attacks on Christmas Island in late February and early March 1942 relations between the British officers and their Indian troops broke down. On the night of 10 March the Indian troops led by a Sikh policemen mutinied killing the five British soldiers and the imprisoning of the remaining 21 Europeans on the island. Later on 31 March, a Japanese fleet arrived at the island and the Indians surrendered.[46]

INA trials

After World War II, the UK tried the commanders of INA as criminals. However, the stories of Indian national army had captured the heart of the citizens and protests spread all over India. Beyond the concurrent campaigns of noncooperation and nonviolent protest, the protest against INA trials spread to include mutinies and wavering support within the British Indian Army. This movement marked the last major campaign in which the forces of the Congress and the Muslim League aligned together; the Congress tricolor and the green flag of the League were flown together at protests. In spite of this aggressive and widespread opposition, the court martial was carried out, and all three defendants were sentenced to deportation for life. This sentence, however, was never carried out, as the immense public pressure of the demonstrations forced Claude Auchinleck, Commander-in-Chief of the Indian Army, to release all three defendants.

During the trial, mutiny broke out in the Royal Indian Navy, incorporating ships and shore establishments of the RIN throughout India, from Karachi to Bombay and from Vizag to Calcutta. The most significant, if disconcerting factor for the Raj, was the significant militant public support that it received. At some places, NCOs in the British Indian Army started ignoring orders from British superiors. In Madras and Pune, the British garrisons had to face revolts within the ranks of the British Indian Army.

Another Army mutiny took place at Jabalpur during the last week of February 1946, soon after the Navy mutiny at Bombay. This was suppressed by force, including the use of the bayonet by British troops. It lasted about two weeks. After the mutiny, about 45 persons were tried by court martial. 41 were sentenced to varying terms of imprisonment or dismissal. In addition, a large number were discharged on administrative grounds. While the participants of the Naval Mutiny were given the freedom fighters' pension, the Jabalpur mutineers got nothing. They even lost their service pension.

Reflecting on the factors that guided the British decision to relinquish the Raj in India, Clement Attlee, the then British prime minister, cited several reasons, the most important of which were the INA activities of Netaji Subhas Chandra Bose, which weakened the Indian Army - the foundation of the British Empire in India- and the RIN Mutiny that made the British realise that the Indian armed forces could no longer be trusted to prop up the Raj. Although Britain had made, at the time of the Cripps' mission in 1942, a commitment to grant dominion status to India after the war this suggests that, contrary to the usual narrative of India's independence struggle, (which generally focuses on Congress and Mahatma Gandhi), the INA and the revolts, mutinies, and public resentment it germinated were an important factor in the complete withdrawal of the Raj from India. Most of the INA. soldiers were set free after cashiering and forfeiture of pay and allowance. On the recommendation of Lord Mountbatten of Burma, and agreed by Nehru, as a precondition for Independence the INA soldiers were not reinducted into the Indian Army. Whether as a measure of the pain that the allies suffered in Imphal and Burma or as an act of vengeance, Mountbatten, Head of Southeast Asia Command, ordered the INA Memorial to its fallen soldiers destroyed when the Singapore was recaptured in 1945. It has been suggested later that Mountbatten's actions may have been to erase completely the records of INA's existence, to prevent the seeds of the idea of a revolutionary socialist liberation force from spreading into the vestiges of its colonies amidst the spectre of cold-war politics already taking shape at the time, and had haunted the Colonial powers before the war. In 1995, the National Heritage Board of Singapore marked the place as a historical site. A Cenotaph has since been erected at the site where the memorial stood.

After the war ended, the story of the INA and the Free India Legion was seen as so inflammatory that, fearing mass revolts and uprisings—not just in India, but across its empire—the British Government forbid the BBC from broadcasting their story. However, the stories of the trials at the Red Fort filtered through. Newspapers reported at the time of the trials that some of the INA soldiers held at Red Fort had been executed, which only succeeded in causing further protests.

RIN Mutiny

The Royal Indian Navy Mutiny (the RIN Mutiny or the Bombay Mutiny) encompasses a total strike and subsequent mutiny by the Indian sailors of the Royal Indian Navy on board ship and shore establishments at Mumbai (Bombay) harbour on 18 February 1946. From the initial flashpoint in Mumbai, the mutiny spread and found support through India, from Karachi to Calcutta and ultimately came to involve 78 ships, 20 shore establishments and 20,000 sailors.

The RIN Mutiny started as a strike by ratings of the Royal Indian Navy on the 18th February in protest against general conditions. The immediate issues of the mutiny were conditions and food, but there were more fundamental matters such as racist behaviour by British officers of the Royal Navy personnel towards Indian sailors, and disciplinary measures being taken against anyone demonstrating pro-nationalist sympathies. By dusk on 19 February, a Naval Central Strike committee was elected. Leading Signalman M.S Khan and Petty Officer Telegraphist Madan Singh were unanimously elected President and Vice-President respectively.[47] The strike found immense support among the Indian population already in grips with the stories of the Indian National Army. The actions of the mutineers were supported by demonstrations which included a one-day general strike in Mumbai, called by the Bolshevik-Leninist Party of India, Ceylon and Burma. The strike spread to other cities, and was joined by the Air Force and local police forces. Naval officers and men began calling themselves the Indian National Navy and offered left-handed salutes to British officers. At some places, NCOs in the British Indian Army ignored and defied orders from British superiors. In Chennai and Pune, the British garrisons had to face revolts within the ranks of the British Indian Army. Widespread rioting took place from Karachi to Calcutta. Famously the ships hoisted three flags tied together — those of the Congress, Muslim League, and the Red Flag of the Communist Party of India (CPI), signifying the unity and demarginalisation of communal issues among the mutineers.

The true judgment of contributions of each of these individual events and revolts to India's eventual independence, and the relative success or failure of each, remains open to historians. Some historians claim that the *Quit India Movement* was ultimately a failure[48] and ascribe more to the destabilisation of the pillar of British power in India the British Indian Armed forces. Certainly the British Prime Minister at the time of Independence, Clement Attlee, deemed the contribution of *Quit India* as minimal, ascribing stupendous importance to the revolts and growing dissatisfaction among Royal Indian Armed Forces as the driving force behind the Raj's decision to leave India[49] [50] Some Indian historians, however, argue that, in fact, it was *Quit India* that succeeded. In support of the latter view, without doubt, the war had sapped a lot of the economic, political and military life-blood of the Empire, and the powerful Indian resistance had shattered the spirit and will of the British government. However, such historians effectively ignore the contributions of the radical movements to transfer of power in 1947. Regardless of whether it was the powerful common call for resistance among Indians that shattered the spirit and will of the British Raj to continue ruling India, or whether it was the ferment of rebellion and resentment among the British Indian Armed Forces[51] [52] what is beyond doubt, is that a population of millions had been motivated as it never had been before to say ultimately that independence was a non-negotiable goal, and every act of defiance and rebel only stoked this fire. In addition, the British people and the British Army seemed unwilling to back a policy of repression in India and other parts of the Empire even as their own country was recovering from war.

Independence and Partition of India, Creation of Pakistan (1947–1950)

On 3 June 1947, Viscount Louis Mountbatten, the last British Governor-General of India, announced the partitioning of the British Indian Empire into India and Pakistan. On 14 August 1947, Pakistan was declared a separate nation from them at 11:57. At 12:02 midnight, on 15 August 1947, India became an independent nation. Violent clashes between Hindus, Sikhs and Muslims followed. Prime Minister Nehru and Deputy Prime Minister Sardar Vallabhbhai Patel invited Mountbatten to continue as Governor General of India. He was replaced in June 1948 by Chakravarti Rajagopalachari. Patel took on the responsibility of unifying 565 princely states, steering efforts by his "iron fist in a velvet glove" policies, exemplified by the use of military force to integrate Junagadh and Hyderabad state (Operation Polo) into India. On the other hand Pandit Jawahar Lal Nehru kept the issue of Kashmir in his hands. The problem of Kashmir has become a permanent cancer in the geographical head of independent India.

The Constituent Assembly completed the work of drafting the constitution on 26 November 1949; on 26 January 1950 the Republic of India was officially proclaimed. The Constituent Assembly elected Dr. Rajendra Prasad as the first President of India, taking over from Governor General Rajgopalachari. Subsequently India annexed Goa and Portugal's other Indian enclaves in 1961), the French ceded Chandernagore in 1951, and Pondicherry and its remaining Indian colonies in 1956, and Sikkim voted to join the Indian Union in 1975.

See also

- Civil resistance
- Goa liberation movement
- Timeline of Indian history

Notes

[1] Chandra 1989, p. 26
[2] Chandra 1989, p. 521
[3] Heehs 1998, p. 9
[4] The English colonial empire, including the territories and trading posts in Asia, came under British control following the union of England and Scotland in 1707.
[5] Heehs 1998, pp. 9–10
[6] Heehs 1998, pp. 11–12
[7] http://lcweb2.loc.gov/cgi-bin/query/r?frd/cstdy:@field(DOCID+bd0018)
[8] Jalal, Ayesha (1994) The Sole Spokesman: Jinnah, the Muslim League and the Demand for Pakistan. Cambridge University Press. ISBN 978-0-521-45850-4
[9] Official website, Government of Pakistan. "The Statesman: Jinnah's differences with the Congress" (http://web.archive.org/web/20060127234847/http://www.pakistan.gov.pk/Quaid/politician7.htm). Archived from the original (http://www.pakistan.gov.pk/Quaid/politician7.htm) on 2006-01-27. . Retrieved 2006-04-20.
[10] Robinson Francis, The Cambridge Encyclopidia of India, 205
[11] Gupta 1997, p. 12
[12] Popplewell 1995, p. 201
[13] Strachan 2001, p. 798
[14] Hoover 1985, p. 252
[15] Brown 1948, p. 300
[16] Strachan 2001, p. 788
[17] Hopkirk 2001, p. 41
[18] Popplewell 1995, p. 234
[19] Tinker 1968, p. 92
[20] Lovett 1920, pp. 94, 187–191
[21] Sarkar 1921, p. 137
[22] Ackerman, Peter, and Duvall, Jack, *A Force More Powerful: A Century of Nonviolent Conflict* p. 74.
[23] Fraser 1977, p. 257
[24] Banglapedia article (http://banglapedia.search.com.bd/HT/J_0130.htm) by Mohammad Shah
[25] *Rowlatt Report* (§109–110); *First Spark of Revolution* by A.C. Guha, pp. 424–34.

[26] N.P. Shankaranarayana Rao "A shfaqulla Khan: The immortal revolutionary" (http://pib.nic.in/feature/feyr2000/fdec2000/f151220001.html) pib.nic.in. Accessed 1 February 2007
[27] Dr. Mahaur Bhagwandas *Kakori Shaheed Smriti* 1977 Lucknow Kakori Shaheed Ardhshatabdi Samaroh Samiti page 30
[28] Sharma Vidyarnav *Yug Ke Devta: Bismil Aur Ashfaq* 2004 Delhi Praveen Prakashan page 118 ISBN 8177830783
[29] 'Krant'M.L.Verma *Sarfaroshi Ki Tamanna* (Part-1) 1997 Delhi Praveen Prakashan page 35
[30] Khaksar Tehrik Ki Jiddo Juhad Volume 1. Author Khaksar Sher Zaman
[31] Weeks, John, *World War II Small Arms*, New York: Galahad Books (1979), ISBN 0-88365-403-2, p. 89
[32] Dr.'Krant'M.L.Verma *Swadhinta Sangram Ke Krantikari Sahitya Ka Itihas* (Vol-2) p.559
[33] "The Congress and The Freedom Movement" (http://web.archive.org/web/20070811001411/http://www.aicc.org.in/the_congress_and_the_freedom_movement.htm#the). Indian National Congress. Archived from the original (http://www.aicc.org.in/the_congress_and_the_freedom_movement.htm#the) on 2007-08-11. . Retrieved 2007-09-24.
[34] Culture and Combat in the Colonies. The Indian Army in the Second World War. Tarak Barkawi. J Contemp History. 41(2), 325–355.pp:332
[35] http://mondediplo.com/2005/05/13wwiiasia
[36] Edwardes, Michael, *The Last Years of British India, Cleveland, World Pub. Co., 1964, p. 93.*

> The Government of India had hoped, by prosecuting members of the INA, to reinforce the morale of the Indian army. It succeeded only in creating unease, in making the soldiers feel slightly ashamed that they themselves had supported the British. If Bose and his men had been on the right side — and all India now confirmed that they were — then Indians in the Indian army must have been on the wrong side. It slowly dawned upon the Government of India that the backbone of the British rule, the Indian army, might now no longer be trustworthy. The ghost of Subhas Bose, like Hamlet's father, walked the battlements of the Red Fort (where the INA soldiers were being tried), and his suddenly amplified figure overawed the conference that was to lead to independence.

[37] *Encyclopedia Britannica*. Indian National army (http://www.britannica.com/ebi/article-9311795). After returning to India the veterans of the INA posed a difficult problem for the British government. The British feared that a public trial for treason on the part of the INA members might embolden anti-British sentiment and erupt into widespread protest and violence. URL Accessed on 19 Aug 06.
[38] Mutinies (http://news.bbc.co.uk/1/hi/world/europe/3684288.stm) (last section).
[39] Many I.N.A. men already executed (http://www.hindustantimes.com/news/specials/Netaji/images/nov_2_45.gif), Lucknow. *The Hindustan Times, 2 November 1945. URL Accessed 11-Aug-06.*
[40] Legacy and assessment of the effects of the mutiny.
[41] Consequences of the I.N.A. Trials
[42] Tribune India (http://www.tribuneindia.com/2006/20060212/spectrum/main2.htm), accessed on 17-Jul-2006
[43] "RIN mutiny gave a jolt to the British" (http://www.tribuneindia.com/2006/20060212/spectrum/main2.htm) by Dhanjaya Bhat, *The Tribune*, 12 February 2006, retrieved 17 July 2006
[44] Majumdar, R.C., Three Phases of India's Struggle for Freedom, Bombay, Bharatiya Vidya Bhavan, 1967, pp. 58–59.
[45] R.C. Majumdar. History of the Freedom Movement in India. ISBN 0-8364-2376-3, reprint. Calcutta, Firma KLM, 1997, vol. III.
[46] http://www.christmas.net.au/history.html
[47] Encyclopaedia of Political Parties. By O.P Ralhan pp1011 ISBN 81-7488-865-9
[48] Banglapædia (http://banglapedia.search.com.bd/HT/Q_0017.htm)
[49] Dhanjaya Bhat, writing in *The Tribune* (http://www.tribuneindia.com/2006/20060212/spectrum/main2.htm), Sunday, 12 February 2006. Spectrum Suppl.
[50] Majumdar, R.C., *Three Phases of India's Struggle for Freedom*, Bombay, Bharatiya Vidya Bhavan, 1967, pp. 58–59.
[51] WWII Asia (http://mondediplo.com/2005/05/13wwiiasia), Le Monde Diplomatique (http://mondediplo.com/), 2005-05-13.
[52] Tribune India (http://www.tribuneindia.com/2006/20060212/spectrum/main2.htm) 2006-02-12.

References

- Library of Congress (http://lcweb2.loc.gov/frd/cs/intoc.html#in0022)
- Forest, G W (2000). *The Indian Mutiny 1857–1858*. Delhi: Low Price Publications. ISBN 81-7536-196-4.
- Cristoforo, Gallolus (1989). *Italian past*. Oxford University Press.
- Nehru, Jawaharlal. *Discovery of India*. Delhi. ISBN 0-19-562359-2.
- Collins, Larry; Lapierre, Dominique (1997). *Freedom at Midnight*. London, England: HarperCollins. ISBN 0006388515.
- Chandra, Bipan; Mridula Mukherjee, Aditya Mukherjee, Sucheta Mahajan, K.N. Panikkar (1989). *India's Struggle for Independence*. New Delhi: Penguin Books. pp. 600. ISBN 9780140107814.
- Heehs, Peter (1998). *India's Freedom Struggle: A Short History*. Delhi: Oxford University Press. pp. 199. ISBN 9780195627985.
- Sarkar, Sumit (1983). *Modern India: 1885-1947*. Madras: Macmillan. pp. 486. ISBN 0333904257.
- Brown, Giles (August 1948). *The Hindu Conspiracy, 1914-1917.The Pacific Historical Review*. **17**. University of California Press. pp. 299–310. ISSN 0030-8684.
- Popplewell, Richard J (1995). *Intelligence and Imperial Defence: British Intelligence and the Defence of the Indian Empire 1904-1924.*. Routledge. ISBN 9780714645803.
- Hoover, Karl. (May 1985). *The Hindu Conspiracy in California, 1913-1918. German Studies Review*. **8**. German Studies Association. pp. 245–261. ISBN 01497952.
- Hopkirk, Peter (1997). *Like Hidden Fire: The Plot to Bring Down the British Empire.*. Kodansha Globe. ISBN 1568361270.
- Fraser, Thomas G (1977). *Germany and Indian Revolution, 1914-18. Journal of Contemporary History, Vol. 12, No. 2 (Apr., 1977), pp. 255-272*. Sage Publications. ISSN 00220094.
- Strachan, Hew (2001). *The First World War. Volume I: To Arms*. Oxford University Press. USA.. ISBN 0199261911.
- Lovett, Sir Verney (1920). *A History of the Indian Nationalist Movement*. New York, Frederick A. Stokes Company. ISBN 8175362499
- Sarkar, B.K. (March 1921). *Political Science Quarterly*. **36**. The Acedemy of Political Science. pp. 136–138. ISSN 00323195.
- Tinker, Hugh (October 1968). *India in the First World War and after.Journal of Contemporary History, 1918-19: From War to Peace*. **3**. Sage Publications. pp. 89–107. ISSN 00220094.
- Dr.'Krant'M.L.Verma *Swadhinta Sangram Ke Krantikari Sahitya Ka Itihas* (3 Volumes) 2006 New Delhi Praveen Prakashan ISBN 8177831224.
- Sharma Vidyarnav *Yug Ke Devta : Bismil Aur Ashfaq* 2004 Delhi Praveen Prakashan ISBN 8177830783.
- 'Krant'M.L.Verma *Sarfaroshi Ki Tamanna* (4 Volumes) 1997 Delhi Praveen Prakashan.
- Dr. Mahaur Bhagwandas *Kakori Shaheed Smriti* 1977 Lucknow Kakori Shaheed Ardhshatabdi Samaroh Samiti.

Further reading

- Majumdar, R.C.. *History of the Freedom movement in India*. ISBN 0836423763.
- Amales Tripathi, Barun De, Bipan Chandra, Freedom Struggle ISBN 81-237-0249-X
- Philip Mason, *A Matter of Honour: An Account of the Indian Army, its Officers and Men*
- Gandhi, Mohandas (1993). *An Autobiography: The Story of My Experiments With Truth*. Boston: Beacon Press. ISBN 0807059099.
- Chandler, Malcolm; Wright, John (2001). *Modern World History*. Heinemann Educational publishers. 2nd Review edition. ISBN 0435311417.

- Collett, Nigel (2007 (New ed)). *The Butcher of Amritsar: General Reginald Dyer*. Hambledon & London. ISBN 1852855754.
- Sofri, Gianni (1995–1999). *Gandhi and India: A Century in Focus*. Janet Sethre Paxia (translator) (English edition translated from the Italian ed.). Gloucestershire: The Windrush Press. ISBN 1900624125.
- Seal, Anil (1968). *Emergence of Indian Nationalism: Competition and Collaboration in the Later Nineteenth Century*. London: Cambridge U.P.. ISBN 0521062748.
- Brown, Judith M., 'Gandhi and Civil Resistance in India, 1917-47', in Adam Roberts and Timothy Garton Ash (eds.), *Civil Resistance and Power Politics: The Experience of Non-violent Action from Gandhi to the Present.* Oxford & New York: Oxford University Press, 2009. ISBN 978-0-19-955201-6.
- Ashman, Sam (December 1997). "India: Imperialism, partition and resistance" (http://pubs.socialistreviewindex.org.uk/isj77/ashman.htm). *International Socialism* (77).

External links

- Muslim freedom martyrs of India (http://twocircles.net/2009oct01/muslim_freedom_martyrs_india.html) - TCN News
- Indian Freedom Fighters (http://www.peopleforever.org/community.aspx?communityid=64) Tribute on PeopleForever.org
- Pakistani Freedom Fighters (http://www.peopleforever.org/community.aspx?communityid=70) Tribute on PeopleForever.org
- Allama Mashriqi (http://allama-mashriqi.8m.com/)
- Origin of Non-Violence (http://www.sikhism.com/articles/origin_non_violence.htm)
- Independence movement (http://www.freeindia.org/dynamic/)
- Mahatma Gandhi (http://web.mahatma.org.in/flash.html)
- Indian Government (http://goidirectory.nic.in/)
- Raja Mahendapratap (http://www.rajamahendrapratap.com)
- Timeline of Indian independence movement (http://www.kamat.com/kalranga/freedom/timeline.htm)
- Truth behind 1857 parts I (http://www.panthic.org/news/129/ARTICLE/2623/2006-07-16.html), II (http://www.panthic.org/news/129/ARTICLE/2639/2006-07-23.html) & III (http://www.panthic.org/news/125/ARTICLE/2660/2006-07-30.html)

Hindustan Construction Company

HCC	
Type	Public Company
Traded as	BSE: 500185 [1] NSE: HCC [2]
Founded	1926
Headquarters	Mumbai, India
Key people	Ajit Gulabchand, CMD Infrastructure Development Company
Revenue	▲ ₹ 7000.185 crore (US$1.56 billion) (2009-2010)[3]
Employees	2600 (2008)
Website	www.hccindia.com [4]

Hindustan Construction Company (HCC, BSE: 500185 [1], NSE: HCC [2]) is a construction company based in Mumbai, India, founded in 1926.[5] HCC works in various sectors including transportation, power, marine, oil and gas pipeline construction, irrigation, utilities and urban infrastructure. HCC specializes in large-scale civil engineering, and developing construction technologies. It was the first construction company in India to be certified for ISO 9001, ISO 14001 and OHSAS 18001 for its quality, environmental and occupational health and safety management systems. The company is headed by Mr. Ajit Gulabchand, Chairman and Managing Director.[6]

Selected projects

Some major projects undertaken by HCC include:

- Bandra-Worli Sea Link, Mumbai[7]
- Ennore Breakwaters, Tamil Nadu
- Kudankulam Nuclear Power Plant, Tamilnadu
- Rajasthan Atomic Power Station, Rajasthan
- Mumbai Pune Expressway
- Godavari Rail Bridge, Andhra Pradesh
- Nathpa Jhakri Hydro-electric Project: Construction of head race tunnel and surge shaft, with Impregilo SPA
- Naini - Allahabad Bridge: One of the first cable-stayed bridges across the River Yamuna, India, with Hyundai Engineering and Construction
- Delhi Metro Project: Underground Metro Corridor with Kumagaigumi Co. Limited, Japan, Skanska AB, Sweden, and Itochu Corporation, Japan,
- East-West Road Corridor Project in Rajasthan, with M/s. Continental Engineering Corporation, Taiwan
- Lavasa hill city, Pune[7]
- 300MW hydroelectric project in Uttarakhand, india.[8]

Disaster Resource Network

HCC is one of the founding members of Disaster Resource Network (DRN),[9] an initiative of the World Economic Forum, formed in the wake of the Gujarat earthquake of 2001. The network acts as a bridge between business and relief organizations during emergencies, to provide relief and rehabilitation support in the mitigation phases of a disaster. Ajit Gulabchand is on the board of DRN Global, and leads DRN India.

On 24th May 2010, HCC completed the repair and restoration of the damaged Malad water supply tunnel. This leads to the drinking water supply to the western suburbs of Mumbai.[10]

External links

- HCC website [4]

References

[1] http://www.bseindia.com/bseplus/StockReach/AdvanceStockReach.aspx?scripcode=500185

[2] http://www.nseindia.com/marketinfo/companyinfo/companysearch.jsp?cons=HCC&section=7

[3] 2010 BSE Data (http://www.bseindia.com/qresann/detailedresult_cons.asp?scrip_cd=500185&qtr=65.5&compname=HINDUSTAN CONSTRUCTION CO.LTD.&quarter=MC2009-2010&checkcons=55c)

[4] http://www.hccindia.com

[5] "HCC Pockets Order Worth Rs 888 Cr From Nuclear Power Corp | TopNews" (http://www.topnews.in/hcc-pockets-order-worth-rs-888-cr-nuclear-power-corp-2262825). Topnews.in. . Retrieved 2010-07-16.

[6] "Wine and wondering with Ajit Gulabchand of Hindustan Construction-The Leisure Lounge-Features-The Economic Times" (http://economictimes.indiatimes.com/features/the-leisure-lounge//articleshow/3173736.cms). Economictimes.indiatimes.com. 2008-06-28. . Retrieved 2010-07-16.

[7] "Hindustan Construction Co. Ltd (HCC), Bandra-Worli Sea, Sonia Gandhi" (http://www.businesswireindia.com/PressRelease.asp?b2mid=19355). Businesswireindia.com. . Retrieved 2010-07-16.

[8] "HCC wins 300MW hydropower project contract from GMR" (http://www.newstatesman.com/energy-and-clean-tech/2010/09/power-project-india-river). New Statesman. . Retrieved 29 September 2010.

[9] ":: India Inc ::" (http://www.india-inc.in/Ajit-Gulabchand.html). India-inc.in. . Retrieved 2010-07-16.

[10] HCC completes Malad Tunnel restoration (http://www.tunnelsonline.info/story.asp?sectioncode=1&storycode=62884&c=2)

Western Ghats

Western Ghats (*Pragadeesh*)	
Sahyadri Mountains	
Range	
 The Western Ghats at Matheran near Mumbai	
Country	India
States	Gujarat, Maharashtra, Goa, Karnataka, Kerala, Tamil Nadu
Cities	Ootacamund, Mahabaleshwar, Madikeri, Munnar
Highest point	Anamudi
- location	Eravikulam, Idukki, South India, Kerala, India
- elevation	2695 m (8842 ft)
Lowest point	Palakkad Gap
- location	Palakkad, Palakkad district, South India, Kerala, India
- elevation	300 m (984 ft)
Length	1600 km (994 mi), N–S
Width	100 km (62 mi), E–W
Area	160000 km^2 (61776 sq mi)
Biome	forests (30%)
Geology	Basalt, Laterite
Period	Cenozoic

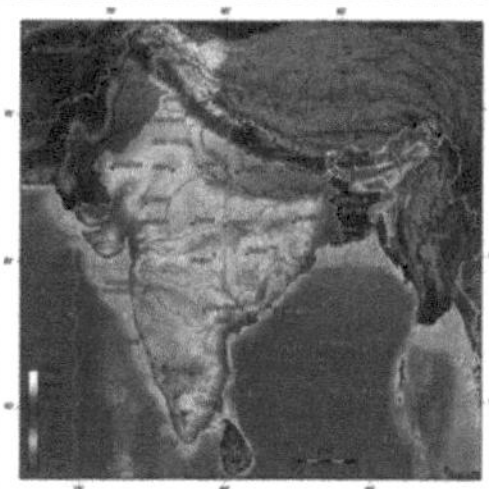

The Western Ghats lie roughly parallel to the west coast of India

The **Western Ghats**, **Western Ghauts** or the **Sahyādri** (Marathi: सह्याद्री, Konkani: सह्याद्री, Kannada: , Tulu: , Malayalam: സഹ്യാദ്രി/സഹ്യപര്വതം, Tamil: மேற்குத் தொடர்ச்சி மலைகள்) is a mountain range along the western side of India. It runs north to south along the western edge of the Deccan Plateau, and separates the plateau from a narrow coastal plain along the Arabian Sea. The Western Ghats block rainfall to the Deccan Plateau. The range starts near the border of Gujarat and Maharashtra, south of the Tapti river, and runs approximately 1600 km through the states of Maharashtra, Goa, Karnataka, Tamil Nadu and Kerala ending at Kanyakumari, at the southern tip of India.

These hills cover 160,000 km² and form the catchment area for complex riverine drainage systems that drain almost 40% of India.[1] The average elevation is around 1200 metres (3900 ft).[2] The area is one of the world's ten "Hottest biodiversity hotspots" and has over 5000 species of flowering plants, 139 mammal species, 508 bird species and 179 amphibian species, many undiscovered species lives in the Western Ghats. At least 325 globally threatened species occur in the Western Ghats.[3]

Geology

The Western Ghats are not true mountains, but are the faulted edge of the Deccan Plateau. They are believed to have been formed during the break-up of the super continent of Gondwana some 150 million years ago. Geophysicists Barron and Harrison from the University of Miami advocate the theory that the west coast of India came into being somewhere around 100 to 80 mya after it broke away from Madagascar. After the break-up, the western coast of India would have appeared as an abrupt cliff some 1,000 meters in height.[4]

Soon after its detachment, the peninsular region of the Indian plate drifted over the Réunion hotspot, a volcanic hotspot in the Earth's lithosphere near the present day location of Réunion. A huge eruption here some 65 mya is thought to have laid down the Deccan Traps, a vast bed of basalt lava that covers parts of central India. These volcanic upthrusts led to the formation of the northern third of the Western Ghats. These dome-shaped uplifts expose underlying 200 mya rocks observed in some parts such as the Nilgiri Hills.[5]

Basalt is the predominant rock found in the hills reaching a depth of 3 km (2 mi). Other rock types found are charnockites, granite gneiss, khondalites, leptynites, metamorphic gneisses with detached occurrences of crystalline limestone, iron ore, dolerites and anorthosites. Residual laterite and bauxite ores are also found in the southern hills.

Mountains

A view of Anamudi, the highest peak of western Ghats (2,695 metres (8,842 ft)) from Eravikulam National Park, Kerala.

The Western Ghats extend from the Satpura Range in the north, go south past Goa, through Karnataka and into Kerala and Tamil Nadu. The major hill range starting from the north is the *Sahyadhri* (*the benevolent mountains*) range. This range is home to many hill stations like Matheran, Lonavala-Khandala, Mahabaleshwar, Panchgani, Amboli Ghat, Kudremukh and Kodagu. The range is called *Sahyadri* in northern Maharashtra and *Sahya Parvatam* in Kerala. The Bili giri rangana Betta southeast of Mysore in Karnataka, meet the Shevaroys (Servarayan range) and Tirumala range farther east, linking the Western Ghats to the Eastern Ghats. In the south the range is known as the *Nilagiri malai* in Tamil Nadu.

Smaller ranges, including the Cardamom Hills and the Nilgiri Hills, are in northwestern Tamil Nadu. The Nilgiri Hills are home to the hill station Ootacamund. In the southern part of the range in the Anaimalai Hills, in western Tamil Nadu and Kerala. Ana Mudi 2695 metres (8842 ft) in Kerala the highest peak in Western Ghats. Chembra Peak 2100 metres (6890 ft), Banasura Peak 2073 metres (6801 ft), Vellarimala 2200 metres (7218 ft) and Agasthya mala 1868 metres (6129 ft) are also in Kerala. Doddabetta is 2637 metres (8652 ft). Mullayanagiri is the highest peak in Karnataka 1950 metres (6398 ft). The Western Ghats in Kerala is home to many tea and coffee plantations. The major gaps in the range are the Goa Gap, between the Maharashtra and Karnataka sections, and the Palghat Gap on the Tamil Nadu/Kerala border between the Nilgiri Hills and the Anaimalai Hills.

Shola Grasslands and forests in the Kudremukh National Park, Western Ghats, Karnataka.

The northern portion of the narrow coastal plain between the Western Ghats and the Arabian Sea is known as the Konkan Coast or simply Konkan, the central portion is called Kanara and the southern portion is called Malabar region or the Malabar Coast. The foothill region east of the Ghats in Maharashtra is known as Desh, while the eastern foothills of the central Karnataka state is known as Malenadu.[6] The largest city within the mountains is the city of Pune (Poona), in the Desh region on the eastern edge of the range. The Biligirirangan Hills lies at the confluence of the Western and Eastern Ghats.

The mountains intercept the rain-bearing westerly monsoon winds, and are consequently an area of high rainfall, particularly on their western side. The dense forests also contribute to the precipitation of the area by acting as a substrate for condensation of moist rising orographic winds from the sea, and releasing much of the moisture back into the air via transpiration, allowing it to later condense and fall again as rain.

Peaks

Salher (Maharashtra), Kalsubai (Maharashtra), Mahableshwar (Maharashtra), Sonsogor (Goa), Kemmangundi (Karnataka), Tadiandamol (Karnataka), Mullayanagiri (Karnataka), (bababudangiri range) (karnataka), Kudremukh (Karnataka), Pushpagiri (Karnataka), Kumara Parvatha (Karnataka), Doddabetta (Tamil Nadu) & Anai Mudi (Kerala).

Lakes and reservoirs

View of western ghatts shows some houses and dwellings

The Western Ghats have several manmade lakes and reservoirs. The well known lakes are the Ooty (2500 m altitude, 34.0 ha) in Nilgiris, and the Kodaikanal (2285 m, 26 ha) and the Berijam in the Palni Hills. The Pookode lake of Wayanad in Kerala at Lakkadi is a beautiful scenic one with boating and garden arrangements. Most of the bigger lakes are situated in the state of Tamil Nadu. Two smaller lakes, the Devikulam (6.0 ha) and the Letchmi Elephant (2.0 ha) are in the Munnar range.

The majority of streams draining the Western Ghats and joining the Rivers Krishna and Kaveri carry water during monsoon months only and have been dammed for hydroelectric and irrigation purposes. The major reservoirs are: Lonavala and Walwahn in Maharashtra; V.V. Sagar, K.R. Sagar and Tungabhadra in the Malenadu area of Karnataka; Mettur, Upper Bhawani, Mukurti, Parson's Valley, Porthumund, Avalanche, Emarold, Pykara, Sandynulla and Glenmorgan in Tamil Nadu; and Kundallay and Maddupatty in the High Range of Kerala. Of these the Lonavla, Walwahn, Upper Bhawani, Mukurti, Parson's Valley, Porthumund, Avalanche, Emarold, Pykara, Sandynulla, Glenmorgan, Kundally and Madupatty are important for their commercial and sport fisheries for trout, mahseer and common carp.[7]

Rivers

The Jog Falls in Karnataka, one of the most spectacular waterfalls in India

The Western Ghats form one of the three watersheds of India, feeding the perennial rivers of India. Important rivers include the Godavari, Krishna, and Kaveri. These rivers flow to the east and drain out into the Bay of Bengal. The west flowing rivers, that drain into the Arabian Sea, are fast-moving, owing to the short distance travelled and steeper gradient. Important rivers include the Mandovi and Zuari. Many of these rivers feed the backwaters of Kerala and Maharashtra. Rivers that flow eastwards of the Ghats drain into the Bay of Bengal. These are comparatively slower moving and eventually merge into larger rivers such as the Kaveri and Krishna. The larger tributaries include the Tunga River, Bhadra river, Bhima River, Malaprabha River, Ghataprabha River, Hemavathi river, Kabini River. In addition there are several smaller rivers such as the Chittar River, Manimuthar River, Kallayi River, Kundali River, Pachaiyar River, Pennar River, Periyar and the Kallayi River.

Fast running rivers and steep slopes have provided sites for many large hydro-electric projects. There are about 50 major dams along the length of the Western Ghats with the earliest project up in 1900 near Khopoli in

Maharashtra.[8] Most notable of these projects are the Koyna Hydroelectric Project in Maharashtra, the Parambikulam Dam in Kerala, and the Linganmakki Dam in Karnataka.[6] The reservoir behind the Koyna Dam, the Shivajisagar Lake, has a length of 50 km (31 mi) and depth of 80 m (262 ft).[9] It is the largest hydroelectric project in Maharashtra, generating 1,920 MW of electric power.[10] Another major Hydro Electric project is Idukki dam in Kerala. This dam is one of the biggest in Asia and generates around 70% of power for Kerala state. Mullai Periyar dam near Thekkady is one of the oldest in the world and a major tourist attractions in Kerala. Water from this dam is drawn to the vast coastal plain of Tamil Nadu, forming a delta and making it rich in vegetation.

Western Ghats in Maharashtra, during the Monsoon

During the monsoon season, numerous streams fed by incessant rain drain off the mountain sides leading to numerous and often spectacular waterfalls. Among the most well known is the Jog Falls, Kunchikal Falls, Sivasamudram Falls, and Unchalli Falls. The Jog Falls is the highest natural plunge waterfall in South Asia and is listed among the 1001 natural wonders of the world.[11] Talakaveri wildlife sanctuary is a critical watershed and the source of the river Kaveri. This region has dense evergreen and semi-evergreen vegetation, with shola-grassland in areas of higher elevation. The steep terrain of the area has resulted in scenic waterfalls along its many mountain streams. Sharavathi and Someshvara Wildlife sanctuaries in Shimoga district are the source of the Tungabhadra River system.The Netravathi river has also its origin at western ghats of India flowing westwards to join Arabian sea at Mangalore.

Climate

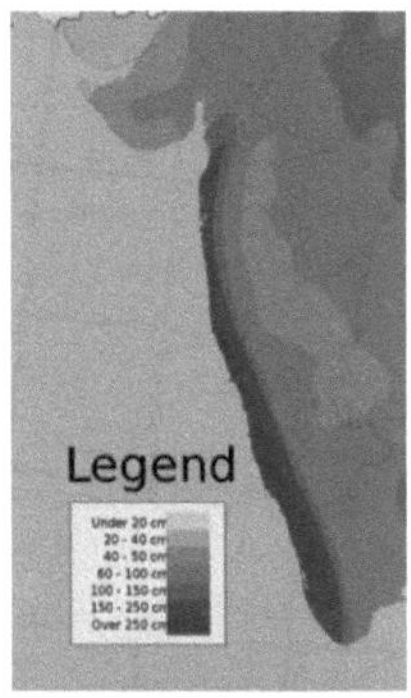

Annual rainfall along the Western Ghat region.

Climate in the Western Ghats varies with altitudinal gradation and distance from the equator. The climate is humid and tropical in the lower reaches tempered by the proximity to the sea. Elevations of 1500 m (4921 ft) and above in the north and 2000 m (6562 ft) and above in the south have a more temperate climate. Average annual temperature here are around 15 °C (60 °F). In some parts frost is common, and temperatures touch the freezing point during the winter months. Mean temperature range from 20 °C (68 °F) in the south to 24 °C (75 °F) in the north. It has also been observed that the coldest periods in the south western ghats coincide with the wettest.[12]

During the monsoon season between June and September, the unbroken Western Ghats chain acts as a barrier to the moisture laden clouds. The heavy, eastward-moving rain-bearing clouds are forced to rise and in the process deposit most of their rain on the windward side. Rainfall in this region averages 3,000–4,000 mm (120–160 in) with localised extremes touching 9,000 mm (350 in). The eastern region of the Western Ghats which lie in the rain shadow, receive far less rainfall averaging about 1,000 mm (40 in) bringing the average rainfall figure to 2,500 mm (150 in). Data from rainfall figures reveal that there is no relationship between the total amount of rain received and the spread of the area. Some areas to the north in Maharashtra while receiving heavier rainfall are followed by long dry spells, while regions closer to the equator receiving less annual rainfall, have rain spells lasting almost the entire year.[12]

Ecoregions

The Western Ghats are home to four tropical and subtropical moist broadleaf forest ecoregions – the North Western Ghats moist deciduous forests, North Western Ghats montane rain forests, South Western Ghats moist deciduous forests, and South Western Ghats montane rain forests.

Nilgiri Hills *(Tectona grandis)* and Wattle.

The northern portion of the range is generally drier than the southern portion, and at lower elevations makes up the North Western Ghats moist deciduous forests ecoregion, with mostly deciduous forests made up predominantly of teak. Above 1,000 meters elevation are the cooler and wetter North Western Ghats montane rain forests, whose evergreen forests are characterized by trees of family *Lauraceae*.

Western Ghats near Rajapalayam

The evergreen Wayanad forests of Kerala mark the transition zone between the northern and southern ecoregions of the Western Ghats. The southern ecoregions are generally wetter and more species-rich. At lower elevations are the South Western Ghats moist deciduous forests, with *Cullenia* the characteristic tree genus, accompanied by teak, dipterocarps, and other trees. The moist forests transition to the drier South Deccan Plateau dry deciduous forests, which lie in its rain shadow to the east.

Western Ghats near Mangalore

Above 1,000 meters are the South Western Ghats montane rain forests, also cooler and wetter than the surrounding lowland forests, and dominated by evergreen trees, although some montane grasslands and stunted forests can be found at the highest elevations. The South Western Ghats montane rain forests are the most species-rich ecoregion in peninsular India; eighty percent of the flowering plant species of the entire Western Ghats range are found in this ecoregion.

Biome Protection

A view of Ponmudi Hills in Thiruvananthapuram district, Kerala.

Historically the Western Ghats were well-covered in dense forests that provided wild foods and natural habitats for native tribal people. Its inaccessibility made it difficult for people from the plains to cultivate the land and build settlements. After the arrival of the British in the area, large swathes of territory were cleared for agricultural plantations and timber. The forest in the Western Ghats has been severely fragmented due to human activities, especially clear felling for tea, coffee, and teak plantations during 1860 to 1950. Species that are rare, endemic and habitat specialists are more adversely affected and tend to be lost faster than other species. Complex and species rich habitats like the tropical rainforest are much more adversely affected than other habitats. [13]

A view of Western Ghats near Pune, India.

The area is ecologically sensitive to development and was declared an ecological hotspot in 1988 through the efforts of ecologist Norman Myers. Though this area covers barely five percent of India's land, 27% of all species of higher plants in India (4,000 of 15,000 species) are found here. Almost 1,800 of these are endemic to the region. The range is home to at least 84 amphibian species, 16 bird species, seven mammals, and 1,600 flowering plants which are not found elsewhere in the world.

The Government of India established many protected areas including 2 biosphere reserves, 13 National parks to restrict human access, several wildlife sanctuaries to protect specific endangered species and many Reserve Forests, which are all managed by the forest departments of their respective state to preserve some of the ecoregions still undeveloped. Many National Parks were initially Wildlife Sanctuaries. The Nilgiri Biosphere Reserve comprising 5500 km² of the evergreen forests of Nagarahole, deciduous forests of Bandipur National Park and Nugu in Karnataka and adjoining regions of Wayanad (വയനാട്) and Mudumalai National Park in the states of Kerala and Tamil Nadu forms the largest contiguous protected area in the Western Ghats.[14] The Western Ghats in Kerala is home to numerous serene hill stations like Munnar (മൂന്നാര്‍), Ponmudi (പൊന്മുടി) and Waynad. The Silent Valley National Park (സൈലന്റ് വാലി) in Kerala is among the last tracts of virgin tropical evergreen forest in India.[15]

Regarding the Western Ghats, in November 2009, Minister of Environment and Forests, Jairam Ramesh said,

> "The Western Ghats has to be made an "ecologically sensitive zone [16]". It is as important as the ecological system of the Himalayas for protection of the environment and climate of the country. The Central government will not give sanction for mining and hydroelectric projects proposed by the State Governments of Maharashtra, Karnataka and Goa that will destroy the Western Ghats eco-system."[17]

In a letter dated June 20, 2009, Mr. Ramesh said,

> "The (proposed) 200-MW Gundia hydel project of Karnataka Power Corporation in Hassan district would drown almost **unknown operator: u','** acres (**unknown operator: u'strong'unknown operator: u',**'km^2) of thick forest in the already endangered Western Ghats along with all its fauna. This is something that both Karnataka and our country can ill-afford." "Power generation should not happen at the cost of ecological security."

The Expert Appraisal Committee appointed by Union Government also said that the project should not be taken up.[18]

A panoramic view of tea plantations in Munnar, Kerala.

World Heritage Site

View of the Western Ghats at Keeriparai, Kanyakumari District, in Tamil Nadu, near the southern end of the range

In 2006, India applied to the UNESCO MAB for the Western Ghats to be listed as a protected World Heritage Site.[19] This will be composed of 7 adjoining areas:

1. Agasthyamalai Sub-Cluster (with Five Site Elements) including: The Agasthyamalai Biosphere Reserve 900 km², includes Kalakkad Mundanthurai Tiger Reserve 806 km², in Tamil Nadu and Neyyar,[20] Peppara[21] and Shendurney[22] Wildlife Sanctuaries and their adjoining areas of Achencoil,[23] Thenmala, Konni,[24] Punalur, Thiruvananthapuram Divisions and Agasthyavanam Special Division in Kerala.[25]
2. Periyar Sub-Cluster (with Six Site Elements) including: Periyar National Park and nature reserve 777 km², in Kerala, Ranni, Konni and Achankovil Forest Divisions. On the eastern side, lying largely in a rain-shadow area with mostly drier forests, lie the Srivilliputtur Wildlife Sanctuary and reserved forests of the Tirunelveli Forest Division.
3. Anamalai Sub-Cluster (with Seven Site Elements) including: Chinnar Wildlife Sanctuary, Eravikulam National Park 90 km², Indira Gandhi National Park, Grass Hills National Park and Karian Shola National Park are located within the larger Indira Gandhi Wildlife Sanctuary 958 km², and Palani Hills National Park 736.87 km² (PRO) in Tamil Nadu and Parambikulam Wildlife Sanctuary 285 km² in Kerala.
4. Nilgiri Sub-Cluster (with Six Site Elements) including: The Nilgiri Biosphere Reserve with Karimpuzha National Park 230 km² (PRO), Silent Valley National Park 89.52 km² and Wayanad Wildlife Sanctuary 344 km² in Kerala, Bandipur National Park 874 km², Mukurthi National Park 78.46 km², Mudumalai National Park 321 km², New Amarambalam Reserved Forest in Tamil Nadu. This sub-cluster constitutes a largely secure forest complex of over 6,000 km², which is one of the most globally significant conservation areas for highly threatened species such as the Asian elephant, tiger and gaur, besides dozens of endangered species in other taxa.
5. Talakaveri Sub-Cluster (with six site elements) including:Brahmagiri Wildlife Sanctuary 181.29 km², Rajiv Gandhi (Nagarhole National Park) 321 km², Pushpagiri Wildlife Sanctuary 92.65 km², Talakaveri Wildlife Sanctuary (105.01 km²) in Karnataka and Aralam Reserved Forest in Kerala.
6. Kudremukh Sub-Cluster (with Five Site Elements) including: Kudremukh National Park 600.32 km², Someshwara Wildlife Sanctuary and surrounding Reserved Forests of Someshwara, Agumbe and Balahalli in karnataka.
7. Sahyadri Sub-Cluster (with Four Site Elements) including: Anshi National Park 340 km², Chandoli National Park 317.67 km², Koyna Wildlife Sanctuary and Radhanagri Wildlife Sanctuary in Maharashtra.

Fauna

The Western Ghats are home to thousands of animal species including at least 325 globally threatened species. Many are endemic species, especially in the amphibian and reptilian classes.

- **Mammals**- There are at least 139 mammal species. A critically endangered mammal of the Western Ghats is the nocturnal Malabar Large-spotted Civet. The arboreal Lion-tailed Macaque is endangered. Only 2500 of this species are remaining.[26] The largest population of Lion tailed macacque is in Silent Valley National Park. Kudremukh National Park also protects a viable population.[27]

  These hill ranges serve as important wildlife corridors, allowing seasonal migration of endangered Asian Elephants. The Nilgiri Bio-sphere is home to the largest population of Asian Elephants and forms an important Project Elephant and Project Tiger reserve. Brahmagiri and Pushpagiri wildlife sanctuaries are important elephant habitats. Karnataka's Ghat areas hold over six thousand elephants (as of 2004) and ten percent of India's critically endangered tiger population.[28]

  The largest population of India's Tigers outside the Sundarbans is in the unbroken forests bordering Karnataka, Tamil Nadu and Kerala. The largest numbers and herds of vulnerable Gaur are found here with the Bandipur National Park and Nagarhole together holding over five thousand Gaur.[29] To the west the forests of Kodagu hold sizeable populations of the endangered Nilgiri Langur.

  Bhadra Wildlife Sanctuary and project tiger reserve in Chikmagalur has large populations of Indian muntjac. Many Asian Elephant, Gaur, Sambar, vulnerable Sloth Bears, Leopard, tiger and Wild Boars dwell in the forests of Karnataka.

  Bannerghatta National Park and Annekal reserve forest is an important elephant corridor connecting the forests of Tamil Nadu with those of Karnataka. Dandeli and Anshi national parks in Uttara Kannada district are home to the Black Panther and normal variety of leopards and significant populations of Great Indian Hornbill. Bhimgad in Belgaum district is a proposed wildlife sanctuary and is home to the endemic critically endangered Wroughton's freetailed bat. the Krishnapur caves close by are one of only three places in the country where the little-known Theobald's tomb bat is found. Large Lesser False Vampire bats are found in the Talevadi caves.[30]

- **Reptiles**- The snake family Uropeltidae of the reptile class is almost entirely restricted to this region.
- **Amphibians**- The amphibians of the Western Ghats are diverse and unique, with more than 80% of the 179 amphibian species being endemic to the region. Most of the endemic species have their distribution in the rainforests of these mountains.[31] The endangered Purple frog was discovered in 2003 to be a living fossil. This species of frog is most closely related to species found in the Seychelles. Four new species of Anurans belonging to the genus Rhacophorus, Polypedates, Philautus and Bufo have been described from the Western Ghats.[32]
- **Fish** – 102 species of fish are listed for the Western Ghats water bodies.[7] Western Ghats streams are home to several brilliantly coloured ornamental fishes like *Red line torpedo barb*, *Red-tailed barb,*[33] *Osteobrama bakeri*, Günther's catfish and freshwater puffer fish *Tetraodon travancoricus*, *Carinotetraodon imitator* and marine forms like *Chelonodon patoca* (Buchanan-Hamilton, 1822);[34] mahseers such as *Malabar mahseer*.[35]
- **Birds**- There are at least 508 bird species. Most of Karnataka's five hundred species of birds[36] are from the Western Ghats region.[37] Bhadra Wildlife Sanctuary is located at the northern end of the Malabar ranges and the southern tip of the Sahyadri ranges and bird species from both ranges can be seen here.

  There are at least 16 species of birds endemic to the western Ghats including the endangered Rufous-breasted Laughingthrush, the vulnerable Nilgiri Wood-pigeon, White-bellied Shortwing and Broad-tailed Grassbird, the near threatened, Grey-breasted Laughingthrush, Black-and-rufous Flycatcher, Nilgiri Flycatcher, and Nilgiri Pipit and the least concern Malabar Parakeet, Malabar Grey Hornbill, White-bellied Treepie, Grey-headed Bulbul, Rufous Babbler, Wynaad Laughingthrush, White-bellied Blue-flycatchers and the Crimson-backed Sunbird.[38]

- **Insects**- There are roughly 6,000 insect species from Kerala alone.[39] Of 334 butterfly species recorded from the Western Ghats, 316 species have been reported from the Nilgiri Biosphere Reserve.[40]
- **Molluscs**- Seasonal rainfall patterns of the Western Ghats necessitate a period of dormancy for its land snails, resulting in their high abundance and diversity including at least 258 species of gastropods from 57 genera and 24 families.[41]

Lion-tailed macaque

Tiger at Bhadra Wildlife Sanctuary

Great Indian Hornbill

Nilgiri Wood-Pigeon

Great Hornbill from Valparai

The endemic land snail *Indrella ampulla*

Malabar Barbet

Flora

The damp forested slopes are the original location of *Piper nigrum*, the black pepper of history and commerce.

See also

- Caecilians of the Western Ghats
- List of Butterflies recorded from the Western Ghats
- Malabar Barbet
- Tropical Rainforest of India

Notes

[1] V.S. Vijayan. "*Research needs for the Western Ghats*" (http://www.westernghatsforum.org/abstract_12_05.pdf). Ashoka Trust for Research in Ecology and the Environment (ATREE). . Retrieved 21 June 2007.

[2] "The Peninsula" (http://web.archive.org/web/20070812064356/http://apmn.icimod.org/mountains/ch2_peninsula.php). *Asia-Pacific Mountain Network* (http://apmn.icimod.org/). Archived from the original (http://apmn.icimod.org/mountains/ch2_peninsula.php) on 12 August 2007. . Retrieved 19 March 2007.

[3] Myers, N., R.A. Mittermeier, C.G. Mittermeier, G.A.B. Da Fonseca, and J. Kent. (2000) "Biodiversity Hotspots for Conservation Priorities." Nature 403:853–858, retrieved 6/1/2007 MYERS, N. (http://www.nature.com/nature/journal/v403/n6772/abs/403853a0.html)

[4] "Paleogeography, 180 million years ago to the present". *Eclogae geologicae Helvetiae* **74** (2): 443–470. 1981.

[5] "Continental drift and the global pattern of sedimentation" (http://www.springerlink.com/content/h93424851308466g/). *Geologische Rundschau* (Berlin / Heidelberg: Springer) **70** (1). February 1981. ISSN 0016-7835 (Print) 1432-1149 (Online). .

[6] "The Geography of India" (http://www.all-about-india.com/Geography-of-India.html). all-aboit-india.com. . Retrieved 19 November 2010.

[7] Sehgal K. L.. "Coldwater fish and fisheries in the Western Ghats, India" (http://www.fao.org/docrep/003/x2614e/x2614e06.htm). FAO. . Retrieved 22 September 2008.

[8] "Indian Dams by River and State" (http://www.rainwaterharvesting.org/resources/statistics/stat233.htm). Rain water harvesting. . Retrieved 19 March 2007.

[9] "Tremors may rock Koyna for another two decade" (http://cities.expressindia.com/fullstory.php?newsid=151253). Indian Express, Pune. 3 October 2005. . Retrieved 19 March 2007.

[10] S. P. Kaushish, B. S. K. Naidu, ed (2002). "Siltation of Reservoirs-Koyna Hydroelectric Project-A Case Study" (http://books.google.com/books?id=iKXDcAAUP5wC&pg=PA1&dq=largest+hydroelectric+project+in+Maharashtra.&ei=o_kbS8nJGJKUNZi2zYQL&client=safari#v=onepage&q=&f=false). *Silting Problems in Hydropower Plants*. Bangkok: Central Board of Irrigation and Power. ISBN 90 5809 238 0. .

[11] Michael Bright, *1001 Natural Wonders of the World* by Barrons Educational Series Inc., Quinted Inc. Publishing, 2005.

[12] R.J. Ranjit Daniels. "Biodiversity of the Western Ghats – An Overview" (http://web.archive.org/web/20070228195706/http://www.wii.gov.in/envis/rain_forest/chapter2.htm). *Wildlife Institute of India* (http://www.wii.gov.in/). Archived from the original (http://www.wii.gov.in/envis/rain_forest/chapter2.htm) on 28 February 2007. . Retrieved 19 March 2007.

[13] Ajith Kumar. "Impact of rainforest fragmentation on small mammals and herpetofauna in the Western Ghats, South India" (http://www.wii.gov.in/publications/researchreports/2002/frag.wii.report.pdf). Salim Centre for Ornithology and Natural History, Coimbatore, India; Ravi Chellam, B. C. Choudhury, Divya Mudappa, Karthikeyan Vasudevan, N. M. Ishwar, Wildlife Institute of India, Dehra Dun, India; Barry Noon, Department of Fish and Wildlife Biology, Colorado State University, Fort Collins, U.S.A., Final Report, April 2002. .

[14] "The Nilgiri Bio-sphere Reserve" (http://cpreec.org/04_phamplets/08_nilgiri_bio_reser/nilgiri_bio_reser.html). .

[15] Elamon Suresh (2006) "*Kerala's Greatest Heritage*", YouTube video, retrieved 4/29/2007 Kerala's Greatest Heritage (http://www.youtube.com/watch?v=X9yEFrhlEBU)

[16] http://envfor.nic.in/legis/eco-senstive.htm

[17] staff (2009-11-21). "No clearance for mining, hydel projects that destroy Western Ghat: Ramesh" (http://beta.thehindu.com/news/states/kerala/article52465.ece). *The Hindu, NEWS » STATES » KERALA* (Palakkad: Kasturi & Sons Ltd). . Retrieved 29 December 2009.

[18] staff (4 August 2009). "Gundia project has not got Centre's nod" (http://www.hindu.com/2009/08/04/stories/2009080456210300.htm). *The hindu* (Chennai: Kasturi & Sons Ltd). . Retrieved 29 December 2009.

[19] "World Heritage sites, Tentative lists, Western Ghats sub cluster" (http://whc.unesco.org/en/tentativelists/2103/). UNESCO, MAB. 2007. . Retrieved 2007-03-30.

[20] Neyyar Wildlife Sanctuary (http://www.keralatourntravel.com/Kerala-WildLife-Neyyar.asp)

[21] Peppara Wildlife Sanctuary (http://www.keralatourntravel.com/Kerala-WildLife-Peppara.asp)

[22] Shendurney Wildlife Sanctuary (http://www.wildkerala.com/html/forest.htm#hendurney)

[23] Achencoil, Kerala (http://www.keralaforest.org/html/tourism/Palaruvi.htm)

[24] Konni, Kerala (http://www.keralaforest.org/html/tourism/konn.htm)

[25] Agasthyavanam Biological Park, Kerala (http://www.keralaforest.org/html/tourism/agastyar.htm)

[26] Participants of CBSG CAMP workshop: Status of South Asian Primates (March 2002) (2004). *Macaca silenus*. 2006. *IUCN Red List of Threatened Species*. IUCN 2006. www.iucnredlist.org (http://www.iucnredlist.org). Retrieved on 9 May 2006. Database entry includes justification for why this species is endangered.

[27] Singh Mewa and Kaumanns Werner (2005) "*Behavioural studies: A necessity for wildlife management*", Current Science, Vol. 89, No. 7, October 10, p.1233. Full text (http://www.ias.ac.in/currsci/oct102005/1230.pdf)

[28] "Karnataka forest department, (forests at a glance -Statistics)" (http://www.karnatakaforest.gov.in/English/forest_glance/forest_at_glance.htm). .

[29] Wildlife populations in Karnataka (http://www.karnatakaforestdepartment.org/English/wild_life_eco_tour/wildlife_eco_tour.htm#overview.)

[30] "Bhimgad awaits government protection" (http://www.sanctuaryasia.com/takeaction/detailcampaign.php?id=33). .

[31] Vasudevan Karthikeyan, *A Report on the Survey of Rainforest Fragments in the Western Ghats for Amphibian Diversity*, retrieved 9/1/2007 (Online version (http://www.wii.gov.in/envis/rain_forest/chapter19.htm)). Introduction

[32] An evaluation of the endemism of the amphibian assemblages from the Western Ghats using molecular techniques Contents & Summary (http://www.wii.gov.in/publications/researchreports/2005/endemism_amphibian_summary_content.pdf)

[33] "?" (http://:www.geocities.com/musammilu/musa.html). geocieties.com. .

[34] "Zoologica" (http://zoologica.lifescience.ntu.edu.tw/Z-10-1/Z-10-1-2.PDF). . Retrieved 18 December 2006.

[35] Silas *et al.*, (2005) Indian Journal of Fisheries, 52(2):125–140

[36] "Karnataka birds" (http://karnatakabirds.net). karnatakabirds.net. .

[37] "Karnataka forest department (forests at a glance – Bio-diversity" (http://www.karnatakaforest.gov.in/English/forest_glance/biodiversity.htm). .

[38] "Restricted-range species" (http://www.birdlife.org/datazone/ebas/index.html?action=EbaHTMDetails.asp&sid=125&m=0). *BirdLife EBA Factsheet 123 Western Ghats*. BirdLife International. 1998. . Retrieved 3 December 2009.

[39] Mathew George and Binoy C.F., *An Overview of Insect Diversity of Western Ghats with Special Reference to Kerala State*, retrieved 7/24/2007. (Online version (http://www.wii.gov.in/envis/rain_forest/chapter3.htm)). Brief summary of work so far carried out on the insect fauna of Western Ghats of Kerala

[40] George Mathew and M. Mahesh Kumar, *State of the Art Knowledge on the Butterflies of Nilgiri Biosphere Reserve*, INDIA, retrieved 9/1/2007 (Online version (http://www.wii.gov.in/envis/rain_forest/chapter6.htm)). Introduction butterfly fauna

[41] Madhyastha N. A., Rajendra, Mavinkurve G. and Shanbhag Sandhya P., *Land Snails of Western Ghats*, retrieved 9/1/2007 (Online version (http://www.wii.gov.in/envis/rain_forest/chapter9.htm)) Introduction Molluscs

References

- A rendezvous with Sahyadri Harshal Mahajan
- Wikramanayake, Eric; Eric Dinerstein; Colby J. Loucks; et al. (2002). *Terrestrial Ecoregions of the Indo-Pacific: a Conservation Assessment.* Island Press; Washington, DC.
- Trek the Sahyadris Harish Kapadia
- Karnataka forest Department- National Parks and WLS (http://www.karnatakaforestdepartment.org/)
- Daniels, R.J. Ranjit, Wildlife institute of India, "Biodiversity in the Western Ghats" (http://www.wii.gov.in/envis/rain_forest/chapter2.htm)
- Ajith Kumar, Sálim Ali Centre for Ornithology and Natural History, Coimbatore, India, Ravi Chellam, B.C.Choudhury, Divya Mudappa, Karthikeyan Vasudevan, N.M.Ishwar, Wildlife Institute of India, Dehra Dun, India, Barry Noon, Department of Fish and Wildlife Biology, Colorado State University, Fort Collins, U.S.A. (2002) "*Impact of Rainforest Fragmentation on Small Mammals and Herpetofauna in the Western Ghats, South India*", Final Report, pp. 146, illus. Full text (http://www.wii.gov.in/faculty/publication/frag.wii.report.pdf) retrieved 3/14/2007
- Verma Desh Deepak (2002) "*Thematic Report on Mountain Ecosystems*", Ministry of Environment and Forests,13pp, retrieved 3/27/2007 Thematic Report on Mountain Ecosystems (http://www.biodiv.org/doc/world/in/in-nr-me-en.doc) Full text, detailed data, not cited.
- Abstracts, Edited by Lalitha Vijayan, Saconr. Vasudeva, University of Dharwad, Priyadarsanan, Atree, Renee Borges, CES, ISSC, Jagdish Krishnaswamy, Atree & WCSP. Pramod, Sacon, Jagannatha Rao, R., FRLHTR. J. Ranjit Daniels, Care Earth, Compiled by S. Somasundaram, Sacon (1–2 December 2005) *Integrating Science and Management of Biodiversity in the Western Ghats* (http://www.westernghatsforum.org/abstract_12_05.pdf), 2nd National Conference of the Western Ghats Forum, Venue: State Forest Service College Coimbatore, Organized by Sálim Ali Centre for Ornithology and Natural History, Anaikatty, Coimbatore – 641108, India. Sponsored by Ministry of Environment and Forests, Government of India. Supported by The Arghyam Foundation, The Ford Foundation & Sir Dorabiji Trust Through Ashoka Trust for Research in Ecology and the Environment (ATREE)
- Shifting Cultivation, Sacred Groves and Conflicts in Colonial Forest Policy in the Western Ghats. (http://google.com/search?q=cache:k5aD-8ZEqdcJ:www.ces.iisc.ernet.in/biodiversity/sahyadri_enews/newsletter/issue30/M%20D%20S%20Chandran.pdf+"+Ankola+High+Forests"&cd=4&hl=en&ct=clnk&gl=ca&client=firefox-a) M,D. Subash Chandran; Chapter 22 PDF ernet.in (http://www.ces.iisc.ernet.in/biodiversity/sahyadri_enews/newsletter/issue30/M D S Chandran.pdf)

External links

- Biodiversityhotspots.org (http://www.biodiversityhotspots.org/xp/hotspots/ghats/Pages/default.aspx)
- Western Gats, India (http://wwf.panda.org/what_we_do/where_we_work/western_ghats/)
- Western Ghats travel guide from Wikitravel

Amusement park

Amusement and **theme parks** are terms for a group of entertainment attractions and rides and other events in a location for the enjoyment of large numbers of people. An amusement park is more elaborate than a simple city park or playground, usually providing attractions meant to cater to children, teenagers, and adults.

Sleeping Beauty Castle at Disneyland

Amusement parks evolved in Europe from fairs and pleasure gardens which were created for people's recreation. The oldest amusement park in the world (opened 1583) is Bakken, at Klampenborg, north of Copenhagen, Denmark. In the United States, world's fairs and expositions were another influence on development of the amusement park industry.[1]

Wonder Mountain at Canada's Wonderland

Most amusement parks have a fixed location, as compared to traveling funfairs and carnivals. These temporary types of amusement parks are usually present for a few days or weeks per year, such as funfairs in the United Kingdom, and carnivals (temporarily set up in a vacant lot or parking lots) and fairs (temporarily operated in a fair ground) in the United States. The temporary nature of these fairs helps to convey the feeling that people are in a different place or time.

Cinderella Castle in Magic Kingdom, Disney World

In common language, **theme park** is often used as a synonym for the term 'amusement park'. A 'theme park' is actually a distinct style of amusement park, for a theme park has landscaping, buildings, and attractions that are based on one or more specific or central themes.[2] [3] A plurality of themes are not required to be considered a 'Theme' park. Despite the long history of amusement parks, where many parks have traditionally incorporated themes into the evolving design and operation of the park, qualifying a park as a theme park, the first park built with the original intension of promoting a specific (or exclusive set of) theme(s), Santa Claus Land (currently known as Holiday World & Splashin' Safari) located in Santa Claus, Indiana, did not open until 1946.[4] [5] Disneyland, located in Anaheim, California, built around the concept of encapsulating multiple theme parks into a single amusement park is often mistakenly noted as the first themed amusement park.

History

Fairs and pleasure gardens

Periodic fairs, such as the Bartholomew Fair which began in England in 1543, are a parent for the modern amusement park. Beginning in the Elizabethan period the fair had evolved into a center of amusement with entertainment, food, games, and carnival-like freak-show attractions.[1] The seasonal celebration was a natural place for development of amusement attractions. Oktoberfest is not only a beer festival but also provided amusement park features beginning in 1810, when the first event was held in Munich, Germany. In the United States, the county and state fairs also played a part in the history of amusement parks. These were annual events that were usually held for a short time, a week or two, to celebrate a good harvest. These fairs featured livestock exhibits, baking and cooking competitions.[6]

Europa-Park, Germany

Canada's Wonderland in Vaughan, Ontario, Canada

Hopi Hari, São Paulo, Brazil

MGM Dizzee World, one of the many amusement parks in Chennai

Vauxhall Gardens, c. 1751.

Amusement parks also grew out of the pleasure gardens that became especially popular at the beginning of the Industrial revolution as an area where one could escape from the grim urban environment. The oldest intact still-surviving amusement park in the world (opened 1583) is Bakken ("The Hill") at Klampenborg, north of Copenhagen, Denmark. The most well known of the parks in London, was Vauxhall Gardens founded in 1661 and closed in 1859.[1] Another long-standing park is Prater in Vienna, Austria, which opened in 1766. This park was conceived as a place where the common person could enjoy a respite in a pastoral setting and participate in the musical culture of the city. Tivoli Gardens, Copenhagen is another example of a European park, dating from 1843, which still exists. These parks consisted of booths, entertainment, fireworks displays and some "rides" such as introduction to the modern railroad. The parks grew to accommodate the expectations of their customers—who were increasingly familiar with the mechanical wonders of industrialization. Rides became a required part of the pleasure garden and by 1896 there were 65 such pleasure parks in London.[1]

First Ferris wheel, World's Columbian Exposition, 1893

Another type of fair is the exposition or world's fair. World's fairs began in 1851 with the construction of the landmark Crystal Palace in London, England. The purpose of the exposition was to celebrate the industrial achievement of the nations of the world (of which Britain just so happened to be the leader).[7] America cities and business saw the world's fair as a way of demonstrating economic and industrial success.[7] People particularly point to the World's Columbian Exposition of 1893 in Chicago, Illinois as an early precursor to the modern amusement park. This fair was an enclosed site that merged entertainment, engineering and education to entertain the masses. It set out to bedazzle the visitors, and successfully did so with a blaze of lights from the "White City." [1] To make sure that the fair was a financial success, the planners included a dedicated amusement concessions area called the Midway Plaisance.[7] Rides from this fair captured imagination of the visitors and of amusement parks around the world, such as the first steel Ferris wheel, which was found in many other amusement areas, such as the Prater by 1896. Also, the experience of the enclosed ideal city with wonder, rides, culture and progress (electricity), was based on the creation of an illusory place. Certainly the precursor of the amusement park experience to come.[1]

The “midway” introduced at the Columbian Exposition would become a standard part of most amusement parks, fairs, carnivals and circuses. The midway contained not only the rides, but other concessions and entertainments such as shooting galleries, penny arcades, games of chance and shows.[6]

Trolley parks, Atlantic City, and Coney Island

In the final decade of the 19th century, the electric trolley lines were developed in most of the larger American cities. Companies that established the trolley lines were directly responsible for establishing amusement parks -- trolley parks -- as destinations of these lines. Trolley parks like Atlanta's Ponce de Leon Park [8], or Reading's Carsonia Park [9] were initially popular natural leisure spots before local streetcar companies purchased the sites, expanding them from picnic groves to include regular entertainments, mechanical amusements, dance halls, sports fields, boats rides, restaurants and other resort facilities. Various sources report the existence of between 1500 and 2000 amusement parks in the United States by 1919.[1]

Steel Pier

Some of these parks were developed in resort locations, such as bathing resorts at the seaside in New Jersey and New York. A premiere example in New Jersey was Atlantic City, a then famed bathing resort. Enterprisers erected amusement parks on piers that extended from the boardwalk out over the ocean. The first of several was Ocean Pier in 1891, followed later by Steel Pier in 1898, both of which boasted rides and attractions typical of that time, such as Midway-style games and electric trolley rides. The boardwalk also had the first Roundabout installed in 1892 by William Somers, a wooden predecessor to the Ferris Wheel. Somers installed two others in Asbury Park and Coney Island. [10] [11] [12]

Other such parks were found along rivers and lakes that provided bathing and water sports such as Riverside Park in Massachusetts, which was founded along the Connecticut River in the 1840s, and Lake Compounce in Connecticut, first established as a bathing beach in 1846.[13]

Another such location was Coney Island in Brooklyn, New York, where a horse-drawn streetcar line brought pleasure seekers to the beach beginning in 1829. In 1875, a million passengers rode the Coney Island Railroad, and in 1876 two million reached Coney Island. Hotels and amusements were built to accommodate both the upper-classes and the working-class. The first carousel was installed in the 1870s, the first roller coaster, the "Switchback Railway", in 1884. It was not until 1895 that the first permanent amusement park in North America opened: Sea Lion Park at Coney Island in Brooklyn. This park was one of the first to charge admission to get into the park in addition to sell tickets for rides within the park.[1]

In 1897, Sea Lion Park was joined by Steeplechase Park, the first of three major amusement parks that would open in the Coney Island area. George Tilyou designed the park to provide thrills and sweep away the restraints of the Victorian crowds. The combination of the nearby population center of New York City and the ease of access to the area made Coney Island the embodiment of the American amusement park.[1] Often, it is Steeplechase Park that comes to mind when one generically thinks of the heyday of Coney Island, but there was also Luna Park (opened in 1903), and Dreamland (opened in 1904). Coney Island was a huge success and by year 1910 attendance on a Sunday could reach a million people.[1] Fueled by the efforts of Frederick Ingersoll, other "Luna Parks" (starting with ones in Pittsburgh and Cleveland in 1905) were quickly erected worldwide and opened to rave reviews.

Fire was a constant threat in those days, as much of the construction within the amusement parks of the era was wooden. In 1911, Dreamland was the first Coney Island amusement park to completely burn down; in 1944, Luna Park also burned to the ground. Most of Ingersoll's Luna Parks were similarly destroyed (usually by arson) before his death in 1927.

The Golden Age

Main entrance to Dorney Park & Wildwater Kingdom, Allentown, Pennsylvania, 1950.

During the Gilded Age, many Americans began working fewer hours[14] and had more disposable income. With new-found money and time to spend on leisure activities, Americans sought new venues for entertainment. Amusement parks, set up outside major cities and in rural areas, emerged to meet this new economic opportunity. These parks reflected the mechanization and efficiency of industrialization while serving as source of fantasy and escape from real life.[1] By the early 1900s, hundreds of amusement parks were operating in the United States and Canada. Trolley parks (established at the end of the trolley line by enterprising streetcar companies) stood outside many cities. Parks like Atlanta's Ponce de Leon[15] and Idora Park[16] near Youngstown, OH took passengers to traditionally popular picnic grounds, which by the late 1890s also often included rides like the Giant Swing, Carousel, and Shoot-the-Chutes. These amusement parks were often based on nationally-known parks or world's fairs: they had names like Coney Island, White City, Luna Park, or Dreamland. The American Gilded Age was, in fact, amusement parks' Golden Age that reigned until the late 1920s.

The Golden Age of amusement parks also included the advent of the kiddie park. Founded in 1925, the original Kiddie Park is located in San Antonio, Texas and is still in operation today. The Kiddie Parks became popular all over America after World War II.[17]

This era saw the development of the new innovations in roller coasters that encouraged extreme drops and speeds to thrill the riders. By the end of the First World War, people seemed to want an even more exciting entertainment, a need met by the roller coasters.[18] Although the development of the automobile provided people with more options for satisfying their entertainment needs, the amusement parks after the war continued to be successful, while urban amusement parks saw declining attendance.[1] The 1920s is more properly known as the Golden Age of roller coasters, being the decade of frenetic building of these rides.[18]

Depression and post-World War II decline

Derelict amusement park in Tanzania, Africa

The Great Depression of the 1930s and World War II during the 1940s saw the decline of the amusement park industry. War saw the affluent urban population move to the suburbs, television became a source of entertainment, and families went to amusement parks less often.[1]

By the 1950s, factors such as urban decay, crime, and even desegregation in the ghettos led to changing patterns in how people chose to spend their free time. Many of the older, traditional amusement parks closed or burned to the ground. Many would be taken out by the wrecking ball to make way for suburban living and development. In 1964, Steeplechase Park, once the king of all amusement parks, closed down for good. The traditional amusement parks which survived, for example, Kennywood, in West Mifflin, Pennsylvania, and Cedar Point, in Sandusky, Ohio, did so in spite of the odds.[1]

The modern amusement park

First parks devoted to a particular theme are precursors for the modern amusement park. The Blackgang Chine amusement park, established in 1843 by Victorian entrepreneur Alexander Dabell, on the Isle of Wight, UK can be considered the oldest existing theme park in the world. The first amusement park on Coney Island, Sea Lion Park was built around a nautical theme.

Modern amusement parks now run differently than those of years past. Amusement parks are usually owned by a large corporate conglomerate which allows capital investment unknown by the traditional family-owned parks. Starting with Disneyland in the 1950s, the park experience became part of a larger package, reflected in a television show, movies, lunch boxes, action figures and finally park rides and costumed characters that make up the "theme." These parks offer an ideal world that serves as an escape from the daily grind. The thrills of the theme parks are often obscured from the outside by high fences or barriers re-enforcing the feeling of escape, they are kept clean and new thrill rides are frequently added to keep people coming back. In addition to this experience, the theme park is either based on a central theme or, divided into several distinctly themed areas, lands or "spaces." Large resorts, such as Walt Disney World in Florida (United States), actually house several different theme parks within their confines.

Today, central Florida and most notably Orlando boasts more theme parks than any other worldwide destination. The northeastern USA region, most notably Pennsylvania, is now a hotbed of traditional surviving amusement parks. In its truest traditional form is Conneaut Lake Park in Conneaut Lake, Pennsylvania. Others include Hersheypark in Hershey, Pennsylvania, Knoebels Groves in Elysburg, Pennsylvania; Kennywood in West Mifflin, Pennsylvania; Idlewild Park in Ligonier, Pennsylvania; Lakemont Park in Altoona, Pennsylvania; Dorney Park & Wildwater Kingdom in Allentown, Pennsylvania; Waldameer Park in Erie, Pennsylvania; and DelGrosso's Amusement Park in Tipton, Pennsylvania.

Educational theme parks

Other parks use outdoor attractions for educational purposes. Epcot center is well known, but there are also Holy Land USA[19] and the Holy Land Experience[20] are theme parks built to inspire Christian piety, Dinosaur World entertains families with dinosaurs in natural settings.

Family-owned theme parks

Narrow gauge mining train going through Calico Ghost Town.

Some theme parks did evolve from more traditional amusement park enterprises, such as Knott's Berry Farm. In the 1920s, Walter Knott and his family sold berries from a roadside stand, which grew to include a restaurant serving fried chicken dinners. Within a few years, lines outside the restaurant were often several hours long. To entertain the waiting crowds, Walter Knott built a Ghost Town in 1940, using buildings relocated from real old west towns such as the Calico, California ghost town and Prescott, Arizona. In 1968, the Knott family fenced the farm, charged admission for the first time, and Knott's Berry Farm officially became an amusement park.[1] Because of its long history, Knott's Berry Farm currently claims to be "America's First Theme Park." Knott's Berry Farm is now owned by Cedar Fair Entertainment Company. Lake Compounce in Bristol, Connecticut may be the true oldest continuously operating amusement park in the United States, open since 1846. Santa Claus Town, which opened in Santa Claus, Indiana in 1935 and included Santa's Candy Castle and other Santa Claus-themed attractions, is considered the first themed attraction in the United States: a pre-cursor to the modern day theme park. Santa Claus Land (renamed Holiday World in 1984) opened in 1946 in Santa Claus, Indiana and many people will argue that it was the first true Theme Park despite Knott's history.[13] In the 1950s the Herschend family took over operation of the tourist attraction, Marvel Cave near Branson, Missouri. Over the next decade they modernized the cave, which led to large numbers of people waiting to take the tour. The Herschend family opened a recreation of the old mining town that once existed atop Marvel Cave. The small village eventually became the theme park, Silver Dollar City. The park is still owned and operated by the Herschends and the family has several other parks including Dollywood, Celebration City and Wild Adventures.

Other theme parks include: Children's Fairyland opened in 1950 in Oakland, California. Another variation of the theme park were the animal theme parks that reintroduced the concept of Sea Lion Park such as Marineland of the Pacific which opened in 1954 which paved the way for SeaWorld parks which eventually added thrill rides.[13]

Disneyland and the corporate-owned park

Walt Disney, however, is often credited with having originated the concept of the themed amusement park, although he was influenced by Knotts Berry Farm owned by Walter Knott (at the time owner of Calico Ghost town) who brought buildings from Calico to increase business at his berry stand located in nearby Buena Park, CA, as well as Tivoli Gardens in Copenhagen. Disney took these influences and melded them with the popular Disney animated characters and his unique vision, and "Disneyland" was born. Disneyland officially opened in Anaheim, California in 1955 and changed the amusement industry forever. Key to the design process of Disney's new park was the replacement of architects with art directors from the film industry.

The years in which Disneyland opened were a sort of stopgap period for the amusement park industry, as many of the older, traditional amusement parks had already closed and many were close to closing their doors. Cedar Point was set to be torn down in the 1950s when local businesspeople were intrigued by the success of Disneyland and saved it from destruction. Other parks were not as lucky, with Steeplechase Park at Coney Island closing in 1964; Riverview Park, Chicago, closed in 1967. Some traditional parks were able to borrow a page from Disneyland and use television to its advantage, such as Kennywood, a park started in 1898 and continuing to operate to the present which used television advertising and featured television personalities at the park.

The first regional theme park, as well as the first Six Flags park, Six Flags over Texas was officially opened in 1961 in Arlington, Texas near Dallas.[21] The first Six Flags theme park was the vision of Angus Wynne, Jr. and helped create the modern, competitive theme park industry. By 1968, the second Six Flags park, Six Flags Over Georgia, opened, and in 1971, Six Flags Over Mid-America (now Six Flags St. Louis) opened near St. Louis, Missouri. Also in 1971 was the opening of the Walt Disney World resort complex in Florida, which is still the largest theme park and resort complex in the world with the Magic Kingdom (1971), Epcot (1982), Disney's Hollywood Studios (1989) and Disney's Animal Kingdom (1998).

During the 1970s, the theme park industry started to mature as a combination of revitalized traditional amusement parks and new ventures funded by larger corporations emerged. Magic Mountain (now a Six Flags park) opened in Valencia, California. Regional parks such as Cedar Point and Kings Island, popular amusement parks in Ohio, moved towards the more modern theme park-concept as well as rotating new roller coasters and modern thrill rides. Also during the mid-1970s, Marriott Corporation built two identical theme parks named "Great America" in northern California and Illinois. The former is now California's Great America and is owned by Cedar Fair, L.P., which now also owns Kings Island and Cedar Point; and the latter is now Six Flags Great America. Many theme parks were hit badly by the Arab oil embargo of 1973 and a number of planned theme parks were scrapped during this time. Most of today's major amusement parks were built in the 1970s.

Perhaps the most indirect evolution of an attraction into a full-fledged theme park is that of Universal Studios Hollywood. Originally just a backlot tram ride tour of the actual studios in Hollywood, the train ride that started in 1964 slowly evolved into a larger attraction with a western stunt show in 1967, "The Parting of the Red Sea" in

1973, a look at props from the movie *Jaws* in 1975, and the *"Conan the Barbarian"* show in 1984. By 1985, the modern era of the Universal Studios Hollywood theme park began with the "King Kong" ride and, in 1990, Universal Studios Florida in Orlando opened. Universal Studios is now the third-largest theme park company in the world, behind Disney and Merlin.

Present and future of amusement parks

Since the 1980s, the amusement park industry has become larger than ever before, with everything from large, worldwide type theme parks such as Disneyworld and Universal Studios Hollywood to smaller and medium-sized theme parks such as the Six Flags parks and countless smaller ventures in many of the states of the U.S. and in countries around the world. Even simpler theme parks directly aimed at smaller children have emerged, including Legoland opened in 1999 in Carlsbad, California (the first Legoland opened in 1968 in Billund, Denmark). The only limit to future theme park ventures is one's imagination.

Dorney Park and Wildwater Kingdom's in Allentown, Pennsylvania

Amusement parks in shopping malls began in the 1990s, blending traditional amusement park entertainments—roller coasters, water parks, carousels, and live entertainment—with hotels, movie theaters, and shopping facilities. Examples of giant mall parks are West Edmonton Mall, Alberta, Canada; Pier 39, San Francisco; Mall of America, Bloomington, Minnesota.

Amusement park owners are also aware of the need to satisfy their aging baby boomer customer base with more restaurants, landscaping, gardens and live entertainment. Kennywood has created in 1995 the "Lost Kennywood" area with classic rides that recall the possibly more tranquil times of the early twentieth century. In 2001, Disney opened the Disney's California Adventure which includes Paradise Pier, a recreation of the traditional seaside amusement park of yesteryear.

Family fun parks starting as miniature golf courses have begun to grow to include batting cages, go-karts, bumper cars, bumper boats and water slides. Some of these parks have grown to include even roller coasters, and traditional amusement parks now also have these competition areas in addition to their thrill rides.

The popularity of theme parks has led to the increase of theming -- "the use of an overarching theme, such as western, to create a holistic and integrated spatial organization of a consumer venue" -- in non-theme park venues. While theme restaurants, casinos, and other themed spaces lack the rides and other features of theme parks, they owe much to the legacy of the theme lands and spatial organization that became popular in theme parks.

For several years, there have been the emergence of a new type of theme parks, in which roller coasters are replaced by shows such as Médinat Alzahra in Tunisia and Puy du Fou in France (dealing with History).

Although domestic visitors still make up around 80 percent of admissions to theme and amusement parks, an aging population in the U.S. and a slowing economy in 2008 are forcing The Walt Disney Company and its competitors to seek their fortunes in emerging tourist markets such as in the Middle East and in China. The Walt Disney Company, accounts for around half of the total industry's revenue in the US as a result of more than 50 million adventure seekers pouring through the gates of its U.S.-based attractions each year.[22]

Admission prices and admission policies

Amusement parks collect much of their revenue from admission fees paid by guests attending the park. Other revenue sources include parking fees, food and beverage sales and souvenirs.

Practically all amusement parks operate using one of two admission principles:

Pay-as-you-go

In this format, a guest enters the park at little or no charge. The guest must then purchase rides individually, either at the attraction's entrance or by purchasing ride tickets (or a similar exchange method, like a token). The cost of the attraction is often based on its complexity or popularity. For example, a guest might pay one ticket to ride a carousel but four tickets to ride a roller coaster. The park may allow guests to purchase unlimited admissions to all attractions within the park. A wristband or pass is then shown at the attraction entrance to gain admission.

Melbourne Luna Park

Disneyland opened in 1955 using the pay-as-you-go format.[23] Initially, guests paid the ride admission fees at the attractions. Within a short time, the problems of handling such large amounts of coins led to the development of a ticket system that, while now out of use, is still part of the amusement-park lexicon.[23] In this new format, guests purchased ticket books that contained a number of tickets, labeled "A," "B" and "C." Rides and attractions using an "A-ticket" were generally simple, with "B-tickets" and "C-tickets" used for the larger, more popular rides. Later, the "D-ticket" was added, then finally the now-famous "E-ticket", which was used on the biggest and most elaborate rides, like Space Mountain. Smaller tickets could be traded up for use on larger rides (i.e., two or three A-tickets would equal a single B-ticket). Disneyland, as well as the Magic Kingdom at Walt Disney World, abandoned this practice in 1982.

The advantages of pay-as-you-go include the following:

- guests pay for only what they choose to experience
- attraction costs can be changed easily to encourage use or capitalize on popularity

The disadvantages of pay-as-you-go include the following:

- guests may get tired of spending money almost continuously
- guests may not spend as much on food or souvenirs

Pay-one-price

An amusement park using the pay-one-price format will charge guests a single, large admission fee. The guest is then entitled to use all or more often almost all of the attractions in the park as often as they wish during their visit. The park usually has some attractions that are not included in the admission charge; these are called "up-charge attractions" and can include bungee jumping or go-kart tracks or games of skill. However, the majority of the park's attractions are included in the admission cost.

The “pay-one-price” ticket was first used by George Tilyou at Steeplechase Park, Coney Island in 1897. The entrance fee was $0.25 for entrance to the 15-acre (61000 m^2) park and visitors could enjoy all of the attractions as much as they wanted.

When Angus Wynne, founder of Six Flags Over Texas, first visited Disneyland in 1959, he noted that park's pay-as-you-go format as a reason to make his park pay-one-price.[24] He thought that a family would be more likely to visit his park if they knew, up front, how much it would cost to attend.[24]

The advantages of pay-one-price include:

- guests can more easily budget their visit
- guests may be more likely to experience an attraction they've already paid for
- lower costs for the park operators, since ticket-takers are not needed at each attractions

The disadvantages of pay-one-price include:

- guests will often be paying for attractions that they do not ride or visit
- guests who are simply coming just to be with their families will have to pay anyway

Today's modern theme parks typically charge a single admission fee for admission and unlimited use of attractions, rides, and shows, where as most modern amusement parks offer free admission yet charge separate fees per attraction.

Rides and attractions

Mechanized thrill machines are what makes an amusement park out of a pastoral, relaxing picnic grove or retreat. Earliest rides include the carousel which was originally developed as a way of practicing and then showing-off expertise at tournament skills such as riding and spearing the ring. By the 19th century, carousels were common in parks around the world. Another such ride which shaped the future of the amusement park was the roller coaster. Beginning as a winter sport in 17th century Russia, these gravity driven railroads were the beginning of the search for even more thrilling amusement park rides. The Columbian Exposition of 1893 was a particular fertile testing ground for amusement rides. The Ferris wheel is the most recognized product of the fair. Many rides are set round a theme.

A park contains a mixture of attractions which can be divided into several categories.

Thrill rides

There is a core set of thrill rides which most amusement parks have, including the enterprise, tilt-a-whirl, the gravitron, chairswing, swinging inverter ship, twister, and the top spin. However, there is constant innovation, with new variations on ways to spin and throw passengers around appearing in an effort to keep attracting customers.

Two rides, on the left the Paratrooper ride and right a tropical trip named Terminator

Roller coasters

Since the late 19th century, amusement parks have featured roller coasters. Roller coasters feature steep drops, sharp curves, and inversions. Roller coasters may be the most attractive aspect of a park, but many people come for other reasons. Amusement parks generally have anywhere from two to seven coasters, depending on space and budget. As of 2011, the record for the most coasters in one park is held by Six Flags Magic Mountain with 18; followed by Cedar Point with 17, Canada's Wonderland with 16 and Kings Island / Kings Dominion with 14.

Roller coasters, such as the Behemoth, at Canada's Wonderland, have fast and steep drops from high altitudes.

Train rides

Amusement park trains have had long and varied history in American amusement parks as well as overseas.

According to various websites and historians, the earliest park trains weren't really trains—they were trolleys. The earliest park trains were mostly custom built. Some of the most common manufacturers were:

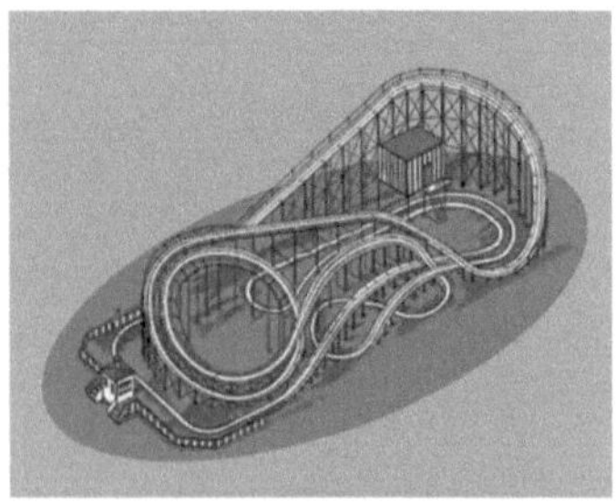

An example of a roller coaster, one of the staples of modern amusement parks

- Allan Herschfield
- Cagney Brothers
- Chance Rides (C.P. Huntington Train)
- Crown Metal Products
- Custom Locomotives
- Miniature Train Co. (MTC)
- The National Amusement Devices Co.(NAD)
- Ottaway
- Sandley
- Tampa Metal Products

Water rides

Amusement parks with water resources generally feature a few water rides, such as the log flume, bumper boats, rapids and rowing boats. Such rides are usually gentler and shorter than roller coasters and many are suitable for all ages. Water rides are especially popular on hot days.

Dark rides

Overlapping with both train rides and water rides, dark rides are enclosed attractions in which patrons travel in guided vehicles along a predetermined path, through an array of illuminated scenes which may include lighting effects, animation, other special effects, music and recorded dialogue.

Transport rides

Transport rides are used to take large amounts of guests from one area in the park to another. They usually cost extra, even in parks where rides are free. They are generally popular as they offer an alternative to walking . Transport rides include chairlifts, monorails, and train rides.

Cuisine

Amusement parks generate a portion of their income through the sale of food and drink to their patrons. Food is routinely sold through food booths, push carts and indoor restaurants. The offerings vary as widely as the amusement parks themselves, and range from common fast food items, like hamburgers, hot dogs, cotton candy, candy apples, donuts and local street foods up to full-service gourmet dishes. Amusement parks with exotic themes may include specialty items or delicacies related to the park's theme. Many restaurants and food stands are operated by the amusement parks themselves, while others are branches of regional or national chains.

Ice-cream and sweets stand at the amusement park at the Louvre, Paris.

Trade associations

- International Association of Amusement Parks and Attractions [25]
- Themed Entertainment Association (USA) [26]
- California Attractions and Parks Association(CAPA) [27]

See also

- Amusement park accidents
- Family Entertainment Center
- Funfair
- Kiddie rides
- List of amusement parks
- List of defunct amusement parks
- Marine mammal park
- Novelty architecture
- Renaissance fair

- World's Fair

References

[1] Adams, Judith A. (1991). *The American Amusement Park Industry: A History of Technology and Thrills*. Boston: Twayne Publishers. ISBN 0805798218.
[2] "Definition of Theme Park" (http://dictionary.reference.com/browse/theme+park). *Dictionary.com*. . Retrieved 2010-09-02.
[3] "Definition of Theme Park" (http://www.merriam-webster.com/dictionary/theme park). *merriam-webster.com*. . Retrieved 2010-09-02.
[4] "A Town Named Santa Claus" (http://www.roadsideamerica.com/story/11443). . Retrieved 2010-09-02.
[5] "History of the Town of Santa Claus, Indiana" (http://www.hohoholdings.com/schistory.htm). . Retrieved 2010-09-02.
[6] Alter, Judy (1997). *Amusement Parks*. New York: Franklin Watts. ISBN 0531203042.
[7] "World's Fairs (1853-1897): A New Idea" (http://www.icewind.net/themepark/History/h_worldfairs.htm). *Midway Plaisance*. . Retrieved 2007-09-29.
[8] http://southernspaces.org/2008/vale-amusements-modernity-technology-and-atlantas-ponce-de-leon-park-1870-1920
[9] http://www.defunctparks.com/parks/PA/carsonia/carsonia.htm
[10] http://explorepahistory.com/hmarker.php?markerId=1096
[11] "New Jersey Online: Atlantic City Museum" (http://www.acmuseum.org/piers.html). Acmuseum.org. . Retrieved 2011-08-10.
[12] Benjamin J. Steinhauser, VersaTrend@gmail.com (1905-01-02). "Timeline" (http://www.cityofatlanticcity.org/con_abo_historytimeline.asp). City of Atlantic City. . Retrieved 2011-08-10.
[13] Samuelson, Dale; Wendy Yegoiants (2001). *The American Amusement Park*. St. Paul, MN: MBI Publishing Company. ISBN 0760309817.
[14] "Eight hours for what we will!" (http://historymatters.gmu.edu/d/6645/). Historymatters.gmu.edu. . Retrieved 2011-08-10.
[15] "Vale of Amusements: Modernity Technology, and Atlanta's Ponce de Leon Park, 1870-1920" (http://southernspaces.org/2008/vale-amusements-modernity-technology-and-atlantas-ponce-de-leon-park-1870-1920). Southernspaces.org. 2008-01-15. . Retrieved 2011-08-10.
[16] "Idora Park - Youngstown, OH" (http://www.defunctparks.com/parks/OH/Idora/idora.htm). Defunctparks.com. . Retrieved 2011-08-10.
[17] "amusement parks" (http://www.kiddiepark.com). Kiddiepark.com. . Retrieved 2011-08-10.
[18] Rutherford, Scott (2000). *The American Roller Coaster*. Osceola, WI: MBI Publishing Company. ISBN 0760306893.
[19] Frances, Chamberlain (November 4, 2001). "The View From/Waterbury; A Hilltop Landmark Undergoes a Revival" (http://query.nytimes.com/gst/fullpage.html?res=9F05E3D71230F937A35752C1A9679C8B63). *The New York Times*. . Retrieved 4 August 2010.
[20] Davis, James D. (2007). "Holyland theme park" (http://www.southflorida.com/travel/sfl-holyland,0,7964224.story). *Sun Sentinel* (Tribune Company). . Retrieved 2009-04-06.
[21] "Sixflags | Six Flags Unofficial Guide" (http://sixflagsmagicmountainguide.com/sixflags.html). Sixflagsmagicmountainguide.com. 2011-04-16. . Retrieved 2011-08-10.
[22] July 2008, Disney & Co., Six Flags, And Other U.S. Theme And Amusement Park Operators Look To Drive Up Revenue With Foreign Visitors (http://www.ibisworld.com/pressrelease/pressrelease.aspx?prid=131:), *IBISWorld*
[23] Bright, Randy (1987). *Disneyland: Inside Story*. Harry N. Abrams. ISBN 978-0810908116.
[24] O'Brien, Tim (1996). *The Essential Guide to Six Flags Theme Parks*. Birmingham, Alabama: Oxmoor House, Inc.. ISBN 0-8487-1247-1.
[25] http://www.iaapa.org
[26] http://www.teaconnect.org
[27] http://www.capalink.org/

Further reading

- Futrell, Jim. *Amusement Parks of New Jersey*. Mechanicsburg, PA: Stackpole Books, 2004.
- Futrell, Jim. *Amusement Parks of New York*. Mechanicsburg, PA: Stackpole Books, 2006.
- Futrell, Jim. *Amusement Parks of Pennsylvania*. Mechanicsburg, PA: Stackpole Books, 2002.
- Futrell, Jim. *Amusement Parks of Virginia, Maryland, and Delaware*. Mechanicsburg, PA: Stackpole Books, 2008.

American Society of Landscape Architects

The **American Society of Landscape Architects** (ASLA) is the national professional association representing landscape architects, with more than 17,000 members in 48 chapters, representing all 50 states, U.S. territories, and 42 countries around the world, plus 68 student chapters. The group was founded on January 4, 1899, to "establish landscape architecture as a recognized profession in North America," "develop educational studies in landscape architecture," and "provide a voice of authority in the 'New Profession'"

One of the Centennial Medallions (1999) that was awarded to Iowa State University for its Central Campus.

ASLA's **unknown operator: u','** square feet (**unknown operator: u'strong'unknown operator: u','**m^2) green roof will be honoured by GRHC in the 'Institutional Intensive' category, which honours innovative green roof design in the US, Canada and Mexico.

Founding members

- Nathan Barrett
- Beatrix Jones Farrand
- Daniel W. Langton
- Charles N. Lowrie
- Warren H. Manning
- Frederick Law Olmsted, Jr.
- John Charles Olmsted (the Society's first president)
- Samuel Parsons, Jr.
- George F. Pentecost, Jr.
- Ossian Cole Simonds
- Downing Vaux (son of Calvert Vaux)

External links

- American Society of Landscape Architects [1]
- *Landscape Architecture* magazine [2]

References

[1] http://www.asla.org/
[2] http://archives.asla.org/nonmembers/lam.html

Initial public offering

An **initial public offering (IPO)**, is the first sale of stock by a formerly private company. It can be used by either small or large companies to raise expansion capital and become publicly traded enterprises. Many companies that undertake an IPO also request the assistance of an Investment Banking firm acting in the capacity of an underwriter to help them correctly assess the value of their shares, that is, the share price (IPO Initial Public Offerings, 2011).

History

In 1602, the Dutch East India Company was the first company in the world to issue stocks and bonds in an initial public offering (Chambers, 2006).

Reasons for listing

When a company lists its securities on a public exchange, the money paid by investors for the newly issued shares goes directly to the company (in contrast to a later trade of shares on the exchange, where the money passes between investors). An IPO, therefore, allows a company to tap a wide pool of investors to provide it with capital for future growth, repayment of debt or working capital. A company selling common shares is never required to repay the capital to investors.

Once a company is listed, it is able to issue additional common shares via a secondary offering, thereby again providing itself with capital for expansion without incurring any debt. This ability to quickly raise large amounts of capital from the market is a key reason many companies seek to go public.

There are several benefits to being a public company, namely:

- Bolstering and diversifying equity base
- Enabling cheaper access to capital
- Exposure, prestige and public image
- Attracting and retaining better management and employees through liquid equity participation
- Facilitating acquisitions
- Creating multiple financing opportunities: equity, convertible debt, cheaper bank loans, etc.
- Increased liquidity for equity holder

Disadvantages of an IPO

There are several disadvantages to completing an initial public offering, namely:

- Significant legal, accounting and marketing costs
- Ongoing requirement to disclose financial and business information
- Meaningful time, effort and attention required of senior management
- Risk that required funding will not be raised
- Public dissemination of information which may be useful to competitors, suppliers and customers

Procedure

IPOs generally involve one or more investment banks known as "underwriters". The company offering its shares, called the "issuer", enters a contract with a lead underwriter to sell its shares to the public. The underwriter then approaches investors with offers to sell these shares.

The sale (allocation and pricing) of shares in an IPO may take several forms. Common methods include:

- Best efforts contract
- Firm commitment contract
- All-or-none contract
- Bought deal
- Dutch auction

A large IPO is usually underwritten by a "syndicate" of investment banks led by one or more major investment banks (lead underwriter). Upon selling the shares, the underwriters keep a commission based on a percentage of the value of the shares sold (called the gross spread). Usually, the lead underwriters, i.e. the underwriters selling the largest proportions of the IPO, take the highest commissions—up to 8% in some cases.

Multinational IPOs may have many syndicates to deal with differing legal requirements in both the issuer's domestic market and other regions. For example, an issuer based in the E.U. may be represented by the main selling syndicate in its domestic market, Europe, in addition to separate syndicates or selling groups for US/Canada and for Asia. Usually, the lead underwriter in the main selling group is also the lead bank in the other selling groups.

Because of the wide array of legal requirements and because it is an expensive process, IPOs typically involve one or more law firms with major practices in securities law, such as the Magic Circle firms of London and the white shoe firms of New York City.

Public offerings are sold to both institutional investors and retail clients of underwriters. A licensed securities salesperson (Registered Representative in the USA and Canada) selling shares of a public offering to his clients is paid a commission from their dealer rather than their client. In cases where the salesperson is the client's advisor it is notable that the financial incentives of the advisor and client are not aligned.

In the US sales can only be made through a final prospectus cleared by the Securities and Exchange Commission.

Investment dealers will often initiate research coverage on companies so their Corporate Finance departments and retail divisions can attract and market new issues.

The issuer usually allows the underwriters an option to increase the size of the offering by up to 15% under certain circumstance known as the greenshoe or overallotment option.

Auction

A venture capitalist named Bill Hambrecht has attempted to devise a method that can reduce the inefficient process. He devised a way to issue shares through a Dutch auction as an attempt to minimize the extreme underpricing that underwriters were nurturing. Underwriters, however, have not taken to this strategy very well which is understandable given that auctions are threatening large fees otherwise payable. Though not the first company to use Dutch auction, Google is one established company that went public through the use of auction. Google's share price rose 17% in its first day of trading despite the auction method. Brokers close to the IPO report that the underwriters actively discouraged institutional investors from buying to reduce demand and send the initial price down. The resulting low share price was then used to "illustrate" that auctions generally don't work.

Perception of IPOs can be controversial. For those who view a successful IPO to be one that raises as much money as possible, the IPO was a total failure. For those who view a successful IPO from the kind of investors that eventually gained from the underpricing, the IPO was a complete success. It's important to note that different sets of investors bid in auctions versus the open market—more institutions bid, fewer private individuals bid. Google may be a special case, however, as many individual investors bought the stock based on long-term valuation shortly after it launched its IPO, driving it beyond institutional valuation.

Pricing

The underpricing of initial public offerings (IPO) has been well documented in different markets (Ibbotson, 1975; Ritter 1984; Levis, 1990; McGuinness, 1992; Drucker and Puri, 2007). While issuers always try to maximize their issue proceeds, the underpricing of IPOs has constituted a serious anomaly in the literature of financial economics. Many financial economists have developed different models to explain the underpricing of IPOs. Some of the models explained it as a consequences of deliberate underpricing by issuers or their agents. In general, smaller issues are observed to be underpriced more than large issues (Ritter, 1984, Ritter, 1991, Levis, 1990)

Historically, some of IPOs both globally and in the United States have been underpriced. The effect of "initial underpricing" an IPO is to generate additional interest in the stock when it first becomes publicly traded. Through flipping, this can lead to significant gains for investors who have been allocated shares of the IPO at the offering price. However, underpricing an IPO results in "money left on the table"—lost capital that could have been raised for the company had the stock been offered at a higher price. One great example of all these factors at play was seen with theglobe.com IPO which helped fuel the IPO mania of the late 90's internet era. Underwritten by Bear Stearns on November 13, 1998, the stock had been priced at $9 per share, and famously jumped 1000% at the opening of trading all the way up to $97, before deflating and closing at $63 after large sell offs from institutions flipping the stock. Although the company did raise about $30 million from the offering it is estimated that with the level of demand for the offering and the volume of trading that took place the company might have left upwards of $200 million on the table.

The danger of overpricing is also an important consideration. If a stock is offered to the public at a higher price than the market will pay, the underwriters may have trouble meeting their commitments to sell shares. Even if they sell all of the issued shares, if the stock falls in value on the first day of trading, it may lose its marketability and hence even more of its value.

Underwriters, therefore, take many factors into consideration when pricing an IPO, and attempt to reach an offering price that is low enough to stimulate interest in the stock, but high enough to raise an adequate amount of capital for the company. The process of determining an optimal price usually involves the underwriters ("syndicate") arranging share purchase commitments from leading institutional investors.

On the other hand, some researchers (e.g. Geoffrey C., and C. Swift, 2009) believe that IPOs are not being under-priced deliberately by issuers and/or underwriters, but the price-rocketing phenomena on issuance days are due to investors' over-reaction (Friesen & Swift, 2009).

Some algorithms to determine underpricing: IPO Underpricing Algorithms

Issue price

A company that is planning an IPO appoints lead managers to help it decide on an appropriate price at which the shares should be issued. There are two ways in which the price of an IPO can be determined: either the company, with the help of its lead managers, fixes a price or the price is arrived at through the process of book building.

Note: Not all IPOs are eligible for delivery settlement through the DTC system, which would then either require the physical delivery of the stock certificates to the clearing agent bank's custodian, or a delivery versus payment (DVP) arrangement with the selling group brokerage firm.

Quiet period

There are two time windows commonly referred to as "quiet periods" during an IPO's history. The first and the one linked above is the period of time following the filing of the company's S-1 but before SEC staff declare the registration statement effective. During this time, issuers, company insiders, analysts, and other parties are legally restricted in their ability to discuss or promote the upcoming IPO (U.S. Securities and Exchange Commission, 2005).

The other "quiet period" refers to a period of 40 calendar days following an IPO's first day of public trading. During this time, insiders and any underwriters involved in the IPO are restricted from issuing any earnings forecasts or research reports for the company. Regulatory changes enacted by the SEC as part of the Global Settlement enlarged the "quiet period" from 25 days to 40 days on July 9, 2002. When the quiet period is over, generally the underwriters will initiate research coverage on the firm. Additionally, the NASD and NYSE have approved a rule mandating a 10-day quiet period after a Secondary Offering and a 15-day quiet period both before and after expiration of a "lock-up agreement" for a securities offering.

Stag profit

Stag profit is a stock market term used to describe a situation before and immediately after a company's Initial public offering (or any new issue of shares). A **stag** is a party or individual who subscribes to the new issue expecting the price of the stock to rise immediately upon the start of trading. Thus, stag profit is the financial gain accumulated by the party or individual resulting from the value of the shares rising.

For example, one might expect a certain I.T. company to do particularly well and purchase a large volume of their stock or shares before flotation on the stock market. Once the price of the shares has risen to a satisfactory level the person will choose to sell their shares and make a stag profit.

Largest IPOs

1. General Motors $23.1 billion (2010)[1]
2. Agricultural Bank of China $22.1 billion (2010)[2]
3. Industrial and Commercial Bank of China $21.9 billion (2006)[3]
4. American International Assurance $20.5 billion (2010)[4]
5. Visa Inc. $19.7 billion (2008)[5]

See also

- Alternative public offering
- Direct public offering
- Equity carve-out
- Mergers and acquisitions (M&A)
- Private placement
- Public offering without listing
- Reverse IPO
- Seasoned equity offering
- SEC Form S-1 (Registration form for certain types of issuers)
- Secondary market offering
- Venture capital

External links

- Initial Public Offering (IPO) Definition and Calendar, Wikinvest
- How IPO works- HowStuff Works [6]
- Most Recent IPO SEC Filings (Form S-1) [7]

References

[1] Bloomberg – GM Says Total Offering Size $23.1 Billion Including Overallotment Options (http://www.bloomberg.com/news/2010-11-26/gm-says-total-offering-size-23-1-billion-including-overallotment-options.html)

[2] Bloomberg – Agricultural Bank of China Sets IPO Record as Size Raised to $22.1 Billion (http://www.bloomberg.com/news/2010-08-15/agricultural-bank-of-china-sets-ipo-record-with-22-1-billion-boosted-sale.html)

[3] Bloomberg – ICBC completed its record $21.9 billion IPO in October 2006 (http://www.bloomberg.com/news/2010-07-28/icbc-to-seek-as-much-as-6-6-billion-in-rights-offer-to-replenish-capital.html)

[4] Bloomberg – AIA's IPO Boosted to $20.5 Billion With Overallotment (http://www.bloomberg.com/news/2010-10-29/aia-s-ipo-boosted-to-20-5-billion-with-overallotment-update1-.html)

[5] WSJ How GM's IPO Stacks Up Against the Biggest IPOs on Record (http://blogs.wsj.com/deals/2010/11/17/how-gms-ipo-stacks-up-against-the-biggest-ipos-on-record/)

[6] http://www.howstuffworks.com/ipo1.htm,

[7] http://www.investorscopes.com/IPOCenter.aspx

- Gregoriou, Greg (2006). *Initial Public Offerings (IPOs)* (http://books.elsevier.com/finance/?isbn=0750679751). Butterworth-Heineman, an imprint of Elsevier. ISBN 0-7506-7975-1.
- Killian, Linda (2006). *IPOs for Everyone* (http://www.renaissancecapital.com/RenCap/AboutUs/Book.aspx). Wiley. ISBN 978-0471399155.
- Emeraldinsight.com (http://www.emeraldinsight.com/Insight/viewContentItem.do?contentType=Article&contentId=1610123), Goergen, M., Khurshed, A. and Mudambi, R. 2007. The Long-run Performance of UK IPOs: Can it be Predicted? *Managerial Finance*, **33**(6): 401–419.
- Bear.cba.ufl.edu (http://bear.cba.ufl.edu/ritter/publ_papers/Why has IPO Underpricing Changed Over Time.pdf) Loughran, T. and Ritter, J.R. 2004. Why Has IPO Underpricing Changed Over Time? *Financial Management*, **33**(3): 5–37.
- RFS.oxfordjournals.org (http://rfs.oxfordjournals.org/cgi/content/abstract/15/2/413) Loughran, T. and Ritter, J.R. 2002. Why Don't Issuers Get Upset About Leaving Money on the Table in IPOs? *Review of Financial Studies*, **15**(2): 413–443.
- Informaworld.com (http://www.informaworld.com/smpp/content~content=a713761351~db=all) Khurshed, A. and Mudambi, R. 2002. The Short Run Price Performance of Investment Trust IPOs on the UK Main Market. *Applied Financial Economics*, **12**(10): 697–706.

- Blackwell-synergy.com (http://www.blackwell-synergy.com/doi/abs/10.1111/1540-6261.00517) Bradley, D.J., Jordan, B.D. and Ritter, J. R. 2003. The Quiet Period Goes Out with a Bang. *Journal of Finance*, **58**(1): 1–36.
- Papers.ssrn.com (http://papers.ssrn.com/sol3/papers.cfm?abstract_id=886408) M.Goergen, M., Khurshed, A. and Mudambi, R. 2006. The Strategy of Going Public: How UK Firms Choose Their Listing Contracts. *Journal of Business Finance and Accounting*, **33**(1&2): 306–328.
- Sciencedirect.com (http://www.sciencedirect.com/science?_ob=ArticleURL&_udi=B6VDH-4D638W4-1&_user=10&_coverDate=07/31/2005&_rdoc=6&_fmt=summary&_orig=browse&_srch=doc-info(#toc#5983#2005#999799995#582452#FLA#display#Volume)&_cdi=5983&_sort=d&_docanchor=&view=c&_ct=6&_acct=C000050221&_version=1&_urlVersion=0&_userid=10&md5=a851cd84c34a896fd87c75690e57057f) Mudambi, R. and Treichel, M.Z. 2005. Cash Crisis in Newly Public Internet-based Firms: An Empirical Analysis. *Journal of Business Venturing*, **20**(4): 543–571.
- Drucker, Steven and M. Puri, 2007, **Banks in Capital Markets**, Handbook of Corporate Finance, Vol. 1, Edited by B. E. Eckbo (2007).
- "IPO Definitions" (http://www.ipoinitialpublicofferings.com/ipo-definitions.htm). IPO Initial Public Offerings. Retrieved 14 September 2011.
- Mondo Visione web site: Chambers, Clem. "Who needs stock exchanges?" (http://www.mondovisione.com/exchanges/handbook-articles/who-needs-stock-exchanges/) *Exchanges Handbook*. Published 2006-07-14. Accessed 21 September 2011
- Friesen, Geoffrey C.; Swift, Christopher. "Overreaction in the thrift IPO aftermarket, Journal of Banking & Finance 33(7)" (http://digitalcommons.unl.edu/financefacpub/5/).
- Anderlini, Jamil (August 13, 2010). "AgBank IPO officially the world's biggest" (http://www.ft.com/cms/s/0/ff7d528c-a6bc-11df-8d1e-00144feabdc0.html?ftcamp=rss). Financial Times. Retrieved 2010-08-13.
- Wines, Michael (July 6, 2010). "China Bank I.P.O. Raises $19 Billion" (http://www.nytimes.com/2010/07/07/business/global/07ipo.html?src=busln). The New York Times. Retrieved 2010-07-06.
- Hu, Bei and Vannucci, Cecile. Bloomberg.com (http://www.bloomberg.com/news/2010-10-28/aia-will-have-hong-kong-trading-debut-today-after-rising-in-gray-market-.html) Published 2010-10-29. Retrieved 2011-09-21
- "Pricing the 'biggest IPO in history'" (http://www.atimes.com/atimes/China_Business/HI29Cb01.html). Published 2006-09-29. Accessed 2011-09-21
- "Quiet Period" (http://www.sec.gov/answers/quiet.htm). Securities and Exchange Commission. August 18, 2005. Retrieved 2008-03-04. "The federal securities laws do not define the term "quiet period," which is also referred to as the "waiting period." However, historically, a quiet period extended from the time a company files a registration statement with the SEC until SEC staff declared the registration statement "effective." During that period, the federal securities laws limited what information a company and related parties can release to the public."

Article Sources and Contributors

Lavasa *Source*: http://en.wikipedia.org/w/index.php?title=Lavasa *Contributors*: 711Techwood, Abdurjaffar, Ak2431989, Akshat.kale, Ananyalal, Angelicam, Ankita1985, Ankitrastogi, Archanajathan, Arjayay, Bearcat, Bkell, Cenarium, Chanvis18, Chris the speller, D6, Devavrat17, Dharmanathf, Drmies, Falastur2, Fallenyangel, Fleetham, Gillyweed, Heading indexing, Imc, J.delanoy, John of Reading, Jpoelma13, LavasaCorp, Lightmouse, LilHelpa, LongLiveMusic, Mandarax, Marokwitz, Minimac, Moonriddengirl, Muhandes, Nshuks7, Phil Bridger, Rait, Rannpháirtí anaithnid, Rich Farmbrough, Rrburke, Rsrikanth05, Salavat, Saturn star, ScottyBerg, Serotta07, Shyamsunder, Stepheng3, Tabletop, TawsifSalam, Utcursch, Verne Equinox, Voyager39, Wikipelli, 182 anonymous edits

Political corruption *Source*: http://en.wikipedia.org/w/index.php?title=Political_corruption *Contributors*: *drew, 2toise, A.t.bruland, Aamonday805, Aawwaajj, Acidify, AdamRetchless, Adrian J. Hunter, Aetheling, Agonzalezb, Ahoerstemeier, Aidenhurleyhaven, Akira625, Alan Liefting, Alansohn, Alci12, Alexplaugh12, Allstarecho, Alphatwo, Andres, Andrzej Kmicic, Andy Marchbanks, Andyjsmith, Angelarhydderch, Angusmclellan, Annawright, Another disinterested reader, Ansori blekethegg, ArglebargleIV, Argon233, Art LaPella, Auric, Avishkargovender, BMF81, Babycakes3ds, Bagdadjenny, Banaticus, Barek, Baron.atreid, Betterusername, Big Adamsky, Bikasuishin, BillO'Slatter, Birdshot9, Blanchardb, Blueprinter, Bobmack89x, Bolivian Unicyclist, BorgQueen, Brekass, Bribeline, Bucoli, CBowers, Caldwell malt, Can't sleep, clown will eat me, Captain panda, Carola56, Catgut, Ccc3, Cekli829, Chad, Charles Matthews, Chris the speller, Confuzion, Coolcaesar, Cpiral, Credema, Criticforaday, Cunado19, Cwescott, D-Rock, D0762, David Biddulph, David.Monniaux, DavidLeeLambert, Dawn Bard, Daytonapost, DeadEyeArrow, Debaserr86, Demmy, Denelson83, DerHexer, Didactik, Dodo von den Bergen, Dondegroovily, DragonflySixtyseven, Drembe, Dthomsen8, Edward, Eeisenberg, Egalitus, Elac310, Ellmist, Elmondo21st, EngineerScotty, Enigmaman, Enimerotita, Enviroboy, EoGuy, Epbr123, Erik9, Eubulides, Everything counts, Explicit, FF2010, Falconleaf, Fastily, Fay.farstad, Felixcatuk, Flyguy649, Forp, Fraggle81, Furrykef, Fæ, GCarty, GD, Gilliam, Ginkgo100, Goethean, Greenwiki, Ground Zero, Guiletti, Gutsul, Gwford, H Padleckas, Hadal, Haemo, Hameed20, HappyInGeneral, Harish india, Haseo9999, Henry Delforn, Hipocrite, Hobartimus, Hodja Nasreddin, Hsxeric, Hu12, Huaiwei, Hugabu, Hugo P, Hydrlix, IFESPF, II MusLiM HyBRiD II, Ibpinfo2011, Iceberg3k, Idont Havaname, ImperatorExercitus, Infrogmation, Innately, Insider-network, Insommia, Iokseng, Ixfd64, J.delanoy, JForget, James086, Jamesbdunn, Jamesmiw, Jamesontai, JanDeFietser, Jeanenawhitney, Jeeves, Jeffhoy, Jiang, Jimduck, Joeblakesley, Joelr31, Jogers, Johann Wolfgang, JohnCD, Jon33, JonHarder, Jonathantreichel, Jongleur100, Jonkerz, Joseph Solis in Australia, Josh Parris, Joy, Jrleighton, Jtuulian, Jusdafax, Jusjih, Jx-10, KF, Kajervi, Katharine908, Kdau, Kevinsproul, Kfitzgib, Kirstyac, Kitty Davis, Kling101, KnowledgeOfSelf, Kozuch, Kummi, Kungfuadam, Kuru, Kurzon, Kwyxz, LarRan, Leftoverz, Legal lamppost, LegitimateAndEvenCompelling, LeilaniLad, Levineps, Lhippocampe, LibLord, Lindmere, Lommer, Lotharsrevenge, Lotje, Lupin, M1ss1ontomars2k4, MCTales, MER-C, MFlet1, Macrakis, Maduromaduro, Malcolm, Marcus22, Marek69, Marekzp, Marrante, Massimomastruzzi, Materialscientist, Matthew Yeager, Mattrix007, Matusz, Maxis ftw, McSly, Mentifisto, Merbabu, Merotoker1, Michael Essmeyer, Miguelzinho, Mikeo, Mikera, Mild Bill Hiccup, Mimihitam, MinervaAB, Mintz0223, Mion, Mission9801, MistyMorn, Modernhiawatha, Modify, Montchav, Moonriddengirl, MrJones, Mystaker1, N1RK4UDSK714, Naddy, Neelix, Neg, Neverquick, Nicmila, Nikjur, Nirbhaesingh, Nk, NoSoftwarePatents, Nsaa, Nscott.odi, Ohnoitsjamie, Olivier, Omicronpersei8, Onlainari, Orange Suede Sofa, OrlinKolev, PFHLai, PL290, PMDrive1061, Paddywhack, Pakfirst, Panwan, Parsecboy, Patstuart, Pavanksingh, Pb30, Pcap, Penbat, Perspicacite, PeterRosello, Pevarnj, PhilKnight, Pierpietro, Pipatron, Plosschen, Politicalheroes, Ptcamn, Qero, Quigley, R'n'B, Radon210, RanEagle, Ravn, Rd232, Rd232 public, RepublicanJacobite, Richard Arthur Norton (1958-), Rjensen, Rjwilmsi, Rkitko, Rlorenc, Roastytoast, Robb hamic, Robertsjursen, Roger4911, Rpyle731, Sachinvenga, Sadads, Sagaciousuk, Saragana, ShadyPolitics, Shanes, Sharpii3, Shenme, Sindhumanne, SiobhanHansa, Sjö, Skupperm, Skysmith, Snoyes, Sonett72, Sonical, Soundofmusicals, Spangineer, Srs, Stanwatch, StephenBuxton, Steven Zhang, Stevertigo, StewartMine, SvenAERTS, TJSwoboda, TRACEint, Tabletop, Taintain, Tango, Teles, Tellyaddict, The Anome, The Magnificent Clean-keeper, The Rambling Man, The Thing That Should Not Be, Thrissel, Tide rolls, Tifgj, Timgould, Tobz1000, Tocharianne, Toparb, TotientDragooned, Tovarischtony, Travelbird, Ultramarine, Universalss, Usyflad10, VJtraveler, Victorgrigas, Vilukatti, Vkvora2001, VolatileChemical, Vranak, Vsion, Vuo, Wadsworth, Walden, Webboc, Weser, Wiki-vr, Wikiuserpune, Wimt, Wireless Keyboard, Wmcg, Wysprgr2005, XB-70, Yonidebest, Yug, Zenwhat, Zvar, Zylu, ﺍﺣﻤﺪ.ﻏﺎﻣﺪﻱ24, 787 anonymous edits

Ministry of Environment and Forests (India) *Source*: http://en.wikipedia.org/w/index.php?title=Ministry_of_Environment_and_Forests_%28India%29 *Contributors*: Atulsnischal, BigrTex, CanisRufus, Chandraguptamaurya, Chaosdruid, Dr. Blofeld, Dwaipayanc, Ekabhishek, EmanWilm, Faisalpcs, Farjad0322, Flatfish, Gueneverey, IndianGeneralist, Marcus334, Neutrality, Nichalp, Pradiptaray, Sachinvenga, Shyamsunder, Veryhuman, Woohookitty, 17 anonymous edits

Hill station *Source*: http://en.wikipedia.org/w/index.php?title=Hill_station *Contributors*: 5 albert square, AI009, AmeriDesi, Andrwsc, Ata Fida Aziz, Azoreg, BACbKA, Black Falcon, BlueHawke, BrainyBabe, Brilliburger, CanadianLinuxUser, Chanakal, Cheong Kok Chun, Chewywiki, Colonies Chris, CommonsDelinker, Criticforaday, D6, Dale Arnett, Dfrg.msc, Dimator, Fatka, Foobaz, Goethean, Gvsdatta, Hamiltondaniel, Harishmm, Herostratus, Hiller coast 3, Hrh80, Hstjw, Imc, IronGargoyle, Joyous!, Jrleighton, Light current, Marcus334, MatthewVanitas, Mattisse, Merbabu, Mervyn, Mhockey, Michael Hardy, Mitsukai, Nabla, NapoliRoma, Nealmcb, Nichalp, Pahari Sahib, Philmcgrove, Plutonium27, Riana, Roland zh, S3000, Sais72, SatuSuro, Scriberius, Shilonite, Shyamsunder, Siddiqui, Skumarla, Spasage, Srikusum, Sstaylor, Stanwatch, Sumanthk, SunKing, Ta bu shi da yu, The Anome, Theo10011, Tlogmer, Tom Radulovich, Tribhuwan, Vegaswikian, Waqas.usman, WikiPedant, Xp54321, YBG, ﻣﺤﺒﻮﺏ ﻋﺎﻟﻢ, 80 anonymous edits

Indian independence movement *Source*: http://en.wikipedia.org/w/index.php?title=Indian_independence_movement *Contributors*: 7, A930913, Aberdonian99, Achyut.kr.de, Acsenray, Addps4cat, AdjustShift, Aeosynth, Aheibam, Ahoerstemeier, Ajaxisgod50, Akhil Bakshi, Alansohn, Aldnonymous, Aleksa Lukic, Alren, Ambuj.Saxena, Andonic, Andrew Gwilliam, Angela, Angusmclellan, Ankur, Anna Lincoln, Aranel, Art LaPella, Arthena, Ashwatham, Astromanity, Atif.hussain, Austriacus, Avaneeshkumarm, AxelBoldt, Baa, Barunroy, Basawala, Bastin, Bcorr, Beaumont, Bender235, Betacommand, Betterusername, Bhadani, Bhawani Gautam, Bijee, Bkell, BlastOButter42, Blink484, BobClive, Bobblewik, Bobo192, Borgx, Brhaspati, Brian0918, Brinlong, Brsaindia, BryanG, Bundelkhandi, C0MRADE, CALR, CBDunkerson, CBM, Caesura, Caitlin000, Callowschoolboy, Caltas, Can't sleep, clown will eat me, Capricorn42, CarTick, Carmichael95, Cbe3000, Cdc, Cfso2325, Chanakal, Chancemill, Charatpratapsingh, Charles Matthews, Chase me ladies, I'm the Cavalry, Chasingsol, Chedda61, Chris 73, Chris the speller, Chrism, Chuck02, Cleared as filed, Cmdrjameson, Cncs wikipedia, CommonsDelinker, Confuzion, Coolguybha, Cremepuff222, Cumulus Clouds, Cyan, DJ Clayworth, DaGizza, Dakutaa, DanMS, Dannyc77, Davemustaine, DavidDouthitt, Davidsteinberg, Dbachmann, Dcrossle, Deepak D'Souza, Deeptrivia, Democraticboy, Deon, Devilicious, Diannaa, Dictabeard, Dimsim37, Dirtyharree, Discospinster, DisillusionedBitterAndKnackered, DocBrown, Domino theory, Dozywalia, Dr.R.K.Chopra, Dr.abhinandan swaroop, Draculadonor, Dspradau, Dudewheresmywallet, Duffyt, Dwaipayanc, Dysprosia, EWikist, Edward, Ekabhishek, Emufarmers, Enkiduo, Epbr123, ErrantX, EscapingLife, Euchiasmus, Evercat, EyeSerene, Fabrictramp, Fadereu, Fbifriday, Fconaway, Fowler&fowler, Friend of the Facts, Frietjes, Funandtrvl, Fundada, Fundamental metric tensor, Fæ, G2bambino, Gaius Cornelius, Ganeshal, Gaurav, Gaurav saraswat, Gaurav1146, Gawaxay, Gensanders, Get-back-world-respect, Ghepeu, GiggityGals, Gilliam, Giridhar.tula, Goethean, Goodnightmush, GraemeLeggett, Graft, Green Giant, Ground Zero, Gurch, Gurubrahma, Hadal, HaeB, HahaBlahBoo, HamburgerRadio, Hamza2007, HarryHenryGebel, Harshavardhan83, Helicaltwang109, Hephaestos, Hetam, Himalayanashoka, Hintha, Hkelkar, Hmains, Homagetocatalonia, Hossain Akhtar Chowdhury, Hoverfish, HwarangGo, IFaqeer, IMpbt, Iamdrugs, Ianwarren, Imc, ImpuMozhi, Indopug, InfectedWithRage, Intelligentsium, InverseHypercube, Iridescent, Irishguy, Ivirivi00, J.delanoy, JEN9841, JaGa, Jackyd101, Jacob.jose, Jaedza, Jagged 85, Jatkins, JayC, Jayantanth, Jcw69, Jiang, Jncraton, John Hill, John K, Jonesjohn02004, Jonoikobangali, Joseph Solis in Australia, Josquius, Jossi, Ka Faraq Gatri, Ka34, Kaihsu, Kashayp123, Kehrbykid, Ken Gallager, Keynes.john.maynard, Kglavin, Khazar, Khoikhoi, King of Hearts, Koavf, Kozuch, Kprateek88, KrakatoaKatie, Krantmlverma, Krishnakoli, Krishnarajv, Kunjan1029, LRBurdak, Leandrod, Lee, Leithp, Lightmouse, LilHelpa, Loopy, Lord of the Isles, Lorenzolamaars, Lostintherush, Luna Santin, MECU, MER-C, Macmanui, Macrakis, Maelnuneb, Magioladitis, Makeemlighter, Mantha.satish, Mar4d, Marek69, Mark.murphy, Martin451, MartinDK, MartinRe, Mattbr, Maxamegalon2000, Maxim, Maximilli, McSly, Mel Etitis, Mera722, Mercurius, Michael Hardy, Mild Bill Hiccup, Miquonranger03, MitaliMehra, MithrandirAgain, Mkamat, Mkill, Modular, Modulatum, Moe Epsilon, Moonriddengirl, Morwen, Mpntod, Mr. Absurd, Msiev, Mtijn, Mudassir Rizwan, Muriel Gottrop, Murtasa, Musaabdulrashid, Mxmeyer, Mithrandir, NAHID, NHSavage, NJW494, NYUboy89, Nauticashades, NeilN, Neilinder, Neutrality, Nichalp, Nifky?, Nightstallion, Nirav.maurya, Nirvana2013, Nlu, Nmpenguin, Nobleeagle, Noor Aalam, Notedgrant, Ocaasi, Ohnoitsjamie, Olivia Curtis, Optimist on the run, Otisjimmy1, Oxymoron83, P-Chan, PCock, Pahari Sahib, Paxse, Pearle, PedroPVZ, Perceval, Percyblakeney, Pharos, Phil Boswell, Philip Trueman, Phoebe, Pkhagah, Pokemaniac493, Prasadsathe, Prashanthns, PrestonH, Prof .R.K.Gupta, Psy wombats, Qxz, R K Chopra, R'n'B, RadioFan, Ragib, Raguks, Rahuloof, Raj86ad, Rak3sh, Rakesh Mehta, Rama's Arrow, Ranam, Randhirreddy, Rastabilly, Ravikiran r, Rbakker99, Rdsmith4, Reddi, Rellis1067, RexNL, Riana, Rich Farmbrough, Richard D. LeCour, Richard Keatinge, Rjwilmsi, Rkondawiki, Rlest, Robby.is.on, Robert1947, Robinfilmmaker, Rohitde, Rokfaith, Ronnotel, Roscelese, Royboycrashfan, Rsloch, Rsrikanth05, Rueben lys, Ruhhhrahul, Rupakbiswas, RxS, RyanCross, S.Bharath, SBC-YPR, SDC, SNIyer12, ST47, Salsa Shark, Sam Korn, Samebchase, Sanchitblazer, Sanfranman59, Saravask, Sardanaphalus, Scotcyclops, Seaphoto, Sebin.baby, Serte, Sesel, Shadowjams, Shahab, Shanel, Shantdey, Shovon76, Shravanshukla, Shreyam, Shyam, Signalhead, Sikandarji, Sikh-history, Skaakt, Skapur, SkyWalker, Snowolf, Snoyes, Socks 01, Soman, Southernstar, Splinter1044, Srimanta.Bhuyan, Stan Shebs, Stationcall, Steven J. Anderson, Storm Rider, Sumanch, SunCreator, Sundar, Sunzlvy, Supten, Sushant gupta, Svlad Jelly, Swami1980, Sze cavalry01, T k, Tabletop, Tamicus, Tarn taran, The Red Hat of Pat Ferrick, The Thing That Should Not Be, Thingg, Thiseye, Threecheersfornick, TimBentley, TimComm, Tjbk tjb, Tommy2010, Tony1, Travisl, TreasuryTag, TriniSocialist, Triplejumper, Twinzor, Ulflarsen, Umar Zulfikar Khan, Umar420e, Utcursch, Varano, Varun Alias Appu, Varungarde, Vegaswikian, Vijaynayer, Vikramssheth, Vinoth22389, WadeSimMiser, Way2ashu, Welsh, Wheaven1, WikHead, WikiPuppies, Wikipedia12345, Wildtornado, Wizardman, World8115, Woudloper, Wrelwser43, Wushi-En, Xaosflux, Yann, Yjfstorehouse, Z1gg1, Zahirahmed, Zakksez, Zhen Lin, Zoezoezoe, Zuggernaut, Zzedar, ~shuri, , , 965 anonymous edits

Hindustan Construction Company *Source*: http://en.wikipedia.org/w/index.php?title=Hindustan_Construction_Company *Contributors*: Aaditya 7, Basar, CBM, Chillum, DGG, Dormskirk, Fieldday-sunday, Hspro71, Htanna, Jonathan Oldenbuck, Jorgesaipeter, Jovianeye, Kuru, KuwarOnline, MightyWarrior, Nshuks7, Prodego, Rajtosh, Sameer.kerkar, Snigbrook, Vipullakers, 29 anonymous edits

Western Ghats *Source*: http://en.wikipedia.org/w/index.php?title=Western_Ghats *Contributors*: Aarem, Abhishek191288, AbsolutDan, Adam.J.W.C., Aditya.sjoshi, Agnisdesigners, Ahoerstemeier, Akash50517, Amarrg, Ameymodak, Anchitk, Anoop.m, Antandrus, Anuragparanjape, Aqarian, Arun11, AshLin, Ashwatham, Asnatu wiki, Astynax, Atulsnischal, Avenue, Binoyjsdk, Blake Burba, Bobblewik, C21K, CNRNair, Can't sleep, clown will eat me, Caroig, Challiyan, Chancemill, Chandrahas.dabke, Charles Matthews, Chinmay153, Chris the speller, Ciiindia, Circeus, Ckatz, CommonsDelinker, Cosmic Engine, Darklilac, Deb, Diannaa, Dineshkannambadi, Dirkbb, Drgovande, Ecoconservant, Ecomantra, Effer, Eggman64, Ekabhishek, El C,

Elockid, Fabartus, Fconaway, Ffirehorse, Fratrep, Frongle, GainLine, Gaius Cornelius, Gauthamprabu, Gene Nygaard, Gnanapiti, Goethean, GoingBatty, Grafen, Hamamelis, Hari, Harishmm, HarshalM, Hasiru, Hegdes, HenryFaber, Hgkamath, Higanesh2003, Howard Alexander, IRP, Imc, Invertzoo, Irarum, J.delanoy, JamesBWatson, Jeroen, Jim Douglas, Jo Weber, Joglekar, John Vandenberg, Jovianeye, Jrsanthosh, Kalyanvarma, Kir360, Kjrajesh, KnightRider, Kungfuadam, Lee2008, Lightmouse, Lokicarbis, Lotje, Marcus334, Matthew Yeager, Mattisse, Melesse, Midgrid, Mountainlion123, Nandish-wild, Nasnema, Nataraja, Neelix, Nichalp, Night w, Nirenraju, Nomadic girl, Notinasnaid, Nv8200p, OJWake, Ogress, PamD, Pancho Fierro, Philip Trueman, Pinecar, Poeloq, Ppntori, PradeepPalat, Pragadeeshraja, Prashanthns, Praveenp, ProveIt, Raghu.kuttan, Raguks, Rajaramraok, Rama's Arrow, Ravirrishi, Red, RedWolf, Rich Farmbrough, Rjwilmsi, Sainath468, Sanfy, Sankaracs, Sardanaphalus, Sathyasaagar, Saurabhkulkarni, Sceptre, Sean.hoyland, Shekhartagra, Sherwin1995, Shweta sinha, Shyamal, Shyamsunder, Siddaarth.s, Slavik262, Slawojarek, SpacemanSpiff, Squids and Chips, Srinivas.girigowda, Ssaa08, Sudarshan Bhat, Sudarshan Bhat Trekker, Tatiraju.rishabh, Tbhotch, Thamps, The Discoverer, Theflyer, Tiddly Tom, Tintin1107, Tom Radulovich, TonySever, Trakesht, Tugo, Tumkurameen, Tvispute, Usandeep, Vaikunda Raja, Venkatraman.blr, Vijaybarve, Vivekmw, Wasbeer, Wbfergus, Wetman, Woohookitty, Wouterhagens, Yoga sawant, Δ, 262 anonymous edits

Amusement park *Source*: http://en.wikipedia.org/w/index.php?title=Amusement_park *Contributors*: *Kat*, 74s181, 777sms, A.arvind.arasu, A3 nm, ACT1, AMuseo, AbsolutDan, Admeta, Aillema, Alansohn, Aldis90, Alexandris, All in, Alton, Alyeska2, Amelie poulain, Amit56, Amitsanyal34, Andrewpmk, Andy, Andycjp, Angie Y., Ansett, Antiussentimnet, Apparition11, Apr1fool, Argetrae, Arteitle, Artemis-Arethusa, Atirage, Axeman89, Bbullot, Bearcat, BeaverWithChainsaw, Beetstra, Betacommand, Bigbadbyte, Bigkirby, Bkell, Bobjgalindo, Bobo192, Bodnotbod, Bogdangiusca, Boing! said Zebedee, Borisblue, BrokenSegue, Butterflyindisguise, Cactus Wren, Calvin 1998, CambridgeBayWeather, Camerong, Capricorn42, Captain-n00dle, Caulde, Ccacsmss, Ceyockey, Chandlerjoeyross, Ckatz, Clarkk, CliffC, Clintville, Cmjc80, Cmputer, Columbia, CommonsDelinker, Contedelavie, Crazypenguin2, CryptoDerk, Crystalline Entity, Cynical, D6, DCGeist, DJSparky, DS1953, DVD R W, DabMachine, Dabomb87, Dale Arnett, Daverees, Daveyjones182, Dblevins2, Dc006f5086, DePiep, Deroiij, DesertSteve, Devin61393, Docu, Dogaroon, Dough4872, Duckdog 2, EVula, Eaolson, Ed g2s, Editk, Egieszl, Egil, Ehurtley, Element4523, Elf, Epbr123, Epolk, Erik, Erik16, Ermeyers, Etf, Everyking, Everythingdaniel, Ewlyahoocom, Excirial, Fanatix, Felisone, Fez2005, Fieldday-sunday, Firsfron, Flowanda, Fluoronaut, FlyingToaster, Foldo, Fourohfour, Frankieroberto, Freedomlinux, Frencheigh, Funandtrvl, Func, Fvw, Gail, George100, Gjs238, Glenn, GrooveDog, Grunners, Gtrmp, Guapodugh, Gunman222, Gwernol, Haham hanuka, Harryboyles, Hektor, Hiram K Hackenbacker, Historian, Hullaballoo Wolfowitz, Ian0503, Iceberg3k, Indiscernible, Infrogmation, IronGargoyle, Isewell, Isnow, Ixfd64, JCam, JCork, JForget, JIP, Jackdude101, Jalo, Janke, Jazmin1980, Jazzmanluc5, Jeffrey Mall, Jeffrey O. Gustafson, JesseW, JoanneB, Joe Kress, John254, JohnCD, Jordanblob, Jordancatalano, Jreconomy, Jsc83, Jusdafax, Just zis Guy, you know?, JzG, KPalicz, Kablammo, Katalaveno, Keegan, Keeves, Ken Gallager, King4057, Kingbuckley, Kjetil r, Kjreilly, Klausok, Kmayen, KnowledgeOfSelf, Krang, Krkr12, Ktr101, Kuru, Kwlkid, LUCPOL, Latitude0116, LeaveSleaves, Legoandy, Legotech, Leo-Roy, Leonard23, Leszek Jańczuk, Levineps, Lightmouse, Liopa, Loranchet, Lorenz kerscher, LtPowers, Lugnad, Lupo, Lyght, M RAD Architecture, MJD86, MKFI, Macrodus, Magistra linguae, Martarius, Martg76, Marxtone, Matthew Mawkes, Maurreen, Mb121, McDoobAU93, Mcpusc, Media lib, Meegs, Melaen, Melloss, Melsaran, Mendaliv, Metron4, Mic, Mikecraig, Mikko-Petteri, Mintguy, Mjrmtg, Montu Man 1011, MorganaFiolett, Mporch, Mr.Z-man, MrOllie, MrSCBaker, Mttcmbs, Murgh, Mvincec, Mwanner, Mxn, Naraht, Natkeeran, NeeluSB, NewGuy4, NickelShoe, Nikpapag, Nikthestoned, Nukeh, OBloodyHell, Obli, Ofelixdacat, Ohnoitsjamie, OlEnglish, Olivier, Orphan Wiki, PA Vacation, PAWiki, Pablojulio123, PatCheng, Patrick, Patricknoddy, Patrol110, Paxse, Peeper, PeteAn, Peter Horn, Peter Isotalo, Pharaoh of the Wizards, Physicistjedi, Piano non troppo, Pig de Wig, Playdagame6991, PoccilScript, Postdlf, Power boxx, PrinceYuki, Quarty, Quill, RCT3man1, RJaguar3, RWardy, Raketooy, RavenRider, RedWolf, Reinoutr, RexNL, Rich Farmbrough, Richard, RickK, Rickyrab, Ridetheory, Rjwilmsi, Rlrcstr100, Robbalvey, Roberto212, Robina Fox, Roke, RoyBoy, Rror, Ryulong, SPUI, SYSS Mouse, Sam Hocevar, Samtheboy, Sciurinæ, Scottie theNerd, Seaphoto, Seaserpent85, Seasonpass Doug, Sebsf, Sfdan, Shadowjams, Shanes, Shootthemagpies, SilverSurferSatrani, Silverpark, SimonArlott, Sir.Roderick, SirLamer, Skaterdude182, Slowking Man, Slulek, Smjg, Solipsist, Sonjaaa, Speedway, Spencer, Spidlock, SpikeJones, StephanKetz, Stevietheman, Stickee, Stuartc185, StudierMalMarburg, SunCreator, Supadawg, SuperJumbo, Superclown, Switchmax, Syp, T0m, TMC1982, Taelornk, Tamiera, TeaDrinker, Tex duped, Tfullwood, The Merciful, TheMindsEye, Thegossip596, Themepark, Themeparkgc, Thore, Thryduulf, TigerShark, Tigerbeat, Timetrial3141592, Tinned Elk, Tinton5, Token2k6, Tommy11111, Tonywalton, Topbanana, Travelguideguru, TravisAF, Traxs7, Tregoweth, Trivialist, TulsaTV, Turbinator, Typhoonchaser, Unclemax27, UrbanNerd, Useight, Utcursch, Valerius Myotis, Vbritto, Vegaswikian, Veinor, Versus22, Vytal, WASD, Waldenbg, WaysToEscape, WereSpielChequers, WhisperToMe, Whoville, Wikipelli, WillMcC, Wmahan, Woohookitty, Xcentaur, Xevious, Xrhetor, Zanimum, Zoicon5, Zotdragon, 753 anonymous edits

American Society of Landscape Architects *Source*: http://en.wikipedia.org/w/index.php?title=American_Society_of_Landscape_Architects *Contributors*: Alexdsouza771, Arbustoo, Astseville, Bratsche, Cburnett, Daderot, Darkcore, Edward Waverley, Lightmouse, Lppa, Malcolma, Poltrackt, Shortride, SimonP, Vrysxy, Wavelength, Wknight94, 8 anonymous edits

Initial public offering *Source*: http://en.wikipedia.org/w/index.php?title=Initial_public_offering *Contributors*: 5464536, 7, Aaron Levy-Forsythe, Ab.er.rant, Ahoerstemeier, Ajelectrowhiz, All-or-none, Alokagga, Altenmann, Andrewpmk, Anticipation of a New Lover's Arrival, The, Art LaPella, Aude, AzadRubel, Azurina, BesselDekker, Bigoperm, Blazotron, Bluezy, Bombastus, Bongwarrior, Bped1985, Brian Willis, Buddha24, CaliforniaAliBaba, Camw, Chancemill, Cometstyles, Conant Webb, Cretog8, DMCer, Dasickis, David Gale, Debiobbi, Deineka, Deiz, Deli nk, Devanium, Dicklyon, Djaconi, Dkorn, DocendoDiscimus, Dogposter, Dougransom, Dritorii, Duedilly, Ed g2s, Edjerge, Emerybob, Epbr123, Eraserhead1, Ewlyahoocom, Faderrattnerb, Feco, Flowanda, Fredbauder, Fredrik, FreplySpang, Gaius Cornelius, Galatee, Galoubet, Gary King, Gnfnrf, Hakseng, Hang Li Po, Heavie, Hellomanwaturname, HenryLi, Hmbr, Hu12, Hvn0413, Ian Pitchford, Ibesmart, Ida Shaw, Ikiatop10, Isa Roarte, J.delanoy, JJ211219, Jarbdpo, Jauerback, Jaysweet, Jerryseinfeld, Jerzy, Jewbacca, Johnteslade, Joshua Issac, Jovany rojo, Jphillips, Julesd, JustAGal, KenBest, Kevin McE, Kku, Krazitrain, Krich, Kurtm3, Lakhotiaatul, Lamro, Law-corrector, Leonid Antonenko, Lightmouse, Lootzyne, Lootzynewiki, Lotje, Lycurgus, MER-C, Manop, Manticore, Martinp23, Maurreen, Mehran, MementoVivere, Meowist, Mic, Michael Hardy, MildInvestor, Minesweeper, Modest Genius, Mordo248, Mr Stephen, Navia2011, Ncmvocalist, NeilN, NickW557, Nopetro, Ohnoitsjamie, Ojay123, Olivier, Onnoysomoy, Pasd, PeepP, Phantomsteve, PhilipO, Piano non troppo, Pikappa, Pippin Bear, Platinum Dragon, Pmlineditor, Poccil, Pparazorback, Ppirozzi, Preschooler.at.heart, RainbowOfLight, Rapsar, Rastrelli F, Rdale, Rich Farmbrough, Richard Arthur Norton (1958-), Richie, Rmudambi, Robinth, Rocket71048576, Rolandograndi, Romanc19s, RoyBoy, SEKIUCHI, Seajaylootz, Sean.hoyland, Seaphoto, Secretlondon, Seekingalpha1980, Sekicho, Shadow1, Shadowjams, SilverStar, Simetrical, Skew-t, Skysmith, Slightsmile, Slowking Man, Smallbones, Sowmya mani, Sphayros, Squirepants101, Sridharan.alwar, St.daniel, Stefancina, Strange Passerby, Sybren, TYelliot, TakuyaMurata, TastyPoutine, Taxman, The Thing That Should Not Be, TheObtuseAngleOfDoom, Theilert, Thomas Paine1776, Thumperward, Tiger888, Tinton5, Tommy2010, Tony1, Tonytonytonyk, Tphalter, Tsuji, Ulric1313, Urbanrenewal, Vamsi.chch, Vegas949, Vegaswikian, Venture Capital in Pakistan, Vipul2606, WJetChao, Wall Street CEO, Wangi, War59312, West.andrew.g, Wikidemon, Wikipediarules2221, Wikky Horse, Wk muriithi, Woohookitty, Xionbox, Yflicker, Zhou Yu, Zoz, , 382 anonymous edits

Image Sources, Licenses and Contributors

file:India Maharashtra location map.svg *Source*: http://en.wikipedia.org/w/index.php?title=File:India_Maharashtra_location_map.svg *License*: unknown *Contributors*: Maximilian Dörrbecker (Chumwa)

file:India Maharashtra locator map.svg *Source*: http://en.wikipedia.org/w/index.php?title=File:India_Maharashtra_locator_map.svg *License*: unknown *Contributors*: Abhijitsathe, Planemad

file:Red pog.svg *Source*: http://en.wikipedia.org/w/index.php?title=File:Red_pog.svg *License*: unknown *Contributors*: Anomie

File:Lavasa Fortune Hotel courtyard, Lavasa, Maharashtra.jpg *Source*: http://en.wikipedia.org/w/index.php?title=File:Lavasa_Fortune_Hotel_courtyard,_Lavasa,_Maharashtra.jpg *License*: unknown *Contributors*: Ankur P from Pune, India

Image:World Map Index of perception of corruption 2010.svg *Source*: http://en.wikipedia.org/w/index.php?title=File:World_Map_Index_of_perception_of_corruption_2010.svg *License*: unknown *Contributors*: User:Talkstosocks

Image:Corrupt-Legislation-Vedder-Highsmith-detail-1.jpeg *Source*: http://en.wikipedia.org/w/index.php?title=File:Corrupt-Legislation-Vedder-Highsmith-detail-1.jpeg *License*: unknown *Contributors*: Artist is Elihu Vedder (1836–1923). Photographed 2007 by Carol Highsmith (1946–), who explicitly placed the photograph in the public domain.

File:India against corruption .png *Source*: http://en.wikipedia.org/w/index.php?title=File:India_against_corruption_.png *License*: unknown *Contributors*: User:Senthil524

Image:Antsirabe - rue principale02.JPG *Source*: http://en.wikipedia.org/w/index.php?title=File:Antsirabe_-_rue_principale02.JPG *License*: unknown *Contributors*: User:Bgag

Image:Ifrane.jpg *Source*: http://en.wikipedia.org/w/index.php?title=File:Ifrane.jpg *License*: unknown *Contributors*: User:Bgag

Image:Valley view.jpg *Source*: http://en.wikipedia.org/w/index.php?title=File:Valley_view.jpg *License*: unknown *Contributors*: Original uploader was Mayur.thakare at en.wikipedia Later version(s) were uploaded by Chakravarthy24x7 at en.wikipedia.

Image:Kalimpongtown.jpg *Source*: http://en.wikipedia.org/w/index.php?title=File:Kalimpongtown.jpg *License*: unknown *Contributors*: Dbenbenn, Nichalp, Shanel

Image:Garut-westjava-indonesia-daytime.jpg *Source*: http://en.wikipedia.org/w/index.php?title=File:Garut-westjava-indonesia-daytime.jpg *License*: unknown *Contributors*: Bennylin, Bgag, 1 anonymous edits

Image:Tea fields (Will Ellis).jpg *Source*: http://en.wikipedia.org/w/index.php?title=File:Tea_fields_(Will_Ellis).jpg *License*: unknown *Contributors*: FlickreviewR, GeorgHH, Gryffindor, Leoboudv, Trialsanderrors, 2 anonymous edits

Image:Baguio panorama.jpg *Source*: http://en.wikipedia.org/w/index.php?title=File:Baguio_panorama.jpg *License*: unknown *Contributors*: Reyrefran, 1 anonymous edits

Image:Sri Lanka Teeplantage.jpg *Source*: http://en.wikipedia.org/w/index.php?title=File:Sri_Lanka_Teeplantage.jpg *License*: unknown *Contributors*: User Wolfgangbeyer on de.wikipedia

Image:Da Lat night 2.jpg *Source*: http://en.wikipedia.org/w/index.php?title=File:Da_Lat_night_2.jpg *License*: unknown *Contributors*: Cheong Kok Chun

Image:Genting4.jpg *Source*: http://en.wikipedia.org/w/index.php?title=File:Genting4.jpg *License*: unknown *Contributors*: user:Andylkl

File:Flag of Madagascar.svg *Source*: http://en.wikipedia.org/w/index.php?title=File:Flag_of_Madagascar.svg *License*: unknown *Contributors*: User:SKopp

File:Flag of Morocco.svg *Source*: http://en.wikipedia.org/w/index.php?title=File:Flag_of_Morocco.svg *License*: unknown *Contributors*: User:Denelson83, User:Zscout370

File:Flag of Myanmar.svg *Source*: http://en.wikipedia.org/w/index.php?title=File:Flag_of_Myanmar.svg *License*: unknown *Contributors*: *drew, AnonMoos, CommonsDelinker, Duduziq, Fry1989, Gunkarta, Homo lupus, Idh0854, Josegeographic, Klemen Kocjancic, Legnaw, Mason Decker, Mattes, Neq00, Nightstallion, Pixeltoo, Rfc1394, SeNeKa, Stevanb, ThomasPusch, UnreifeKirsche, Vividuppers, WikipediaMaster, Xiengyod, Zscout370, , 9 anonymous edits

File:Flag of Cambodia.svg *Source*: http://en.wikipedia.org/w/index.php?title=File:Flag_of_Cambodia.svg *License*: unknown *Contributors*: User:Nightstallion

File:Flag of Indonesia.svg *Source*: http://en.wikipedia.org/w/index.php?title=File:Flag_of_Indonesia.svg *License*: unknown *Contributors*: User:Gabbe, User:SKopp

File:Flag of Malaysia.svg *Source*: http://en.wikipedia.org/w/index.php?title=File:Flag_of_Malaysia.svg *License*: unknown *Contributors*: User:SKopp

File:Flag of Pakistan.svg *Source*: http://en.wikipedia.org/w/index.php?title=File:Flag_of_Pakistan.svg *License*: unknown *Contributors*: Abaezriv, AnonMoos, Badseed, Dbenbenn, Duduziq, F. F. Fjodor, Fry1989, Gabbe, Himasaram, Homo lupus, Juiced lemon, Katarighe, Klemen Kocjancic, Mattes, Minna Sora no Shita, Mollajutt, Neq00, Pumbaa80, Rfc1394, Srtxg, TFCforever, ThomasPusch, Túrelio, Zscout370, 9 anonymous edits

File:Flag of the Philippines.svg *Source*: http://en.wikipedia.org/w/index.php?title=File:Flag_of_the_Philippines.svg *License*: unknown *Contributors*: Aira Cutamora

File:Flag of Sri Lanka.svg *Source*: http://en.wikipedia.org/w/index.php?title=File:Flag_of_Sri_Lanka.svg *License*: unknown *Contributors*: Zscout370

File:Flag of Vietnam.svg *Source*: http://en.wikipedia.org/w/index.php?title=File:Flag_of_Vietnam.svg *License*: unknown *Contributors*: user:Lưu Ly

File:Flag of Australia.svg *Source*: http://en.wikipedia.org/w/index.php?title=File:Flag_of_Australia.svg *License*: unknown *Contributors*: Anomie, Mifter

File:British_Indian_Empire_1909_Imperial_Gazetteer_of_India.jpg *Source*: http://en.wikipedia.org/w/index.php?title=File:British_Indian_Empire_1909_Imperial_Gazetteer_of_India.jpg *License*: unknown *Contributors*: Edinburgh Geographical Institute; J. G. Bartholomew and Sons.

Image:Clive.jpg *Source*: http://en.wikipedia.org/w/index.php?title=File:Clive.jpg *License*: unknown *Contributors*: Ashrf1979, Dancingwombatsrule, Flominator, Jappalang, Jungpionier, Madmedea, Nikkimaria, Podzemnik, Ranveig, Thuresson, 4 anonymous edits

Image:Indian revolt of 1857 states map.svg *Source*: http://en.wikipedia.org/w/index.php?title=File:Indian_revolt_of_1857_states_map.svg *License*: unknown *Contributors*: Abhishekjoshi, Juliancolton, Planemad, Roland zh, Wknight94, Zykasaa, 3 anonymous edits

Image:Image-Secundra Bagh after Indian Mutiny higher res.jpg *Source*: http://en.wikipedia.org/w/index.php?title=File:Image-Secundra_Bagh_after_Indian_Mutiny_higher_res.jpg *License*: unknown *Contributors*: Nauticashades, Pinkville, Roland zh, Romary, Victuallers, Wst, 2 anonymous edits

File:Bal G. Tilak.jpg *Source*: http://en.wikipedia.org/w/index.php?title=File:Bal_G._Tilak.jpg *License*: unknown *Contributors*: Arianna, Innotata, Kensplanet, Mu

Image:Majinnah4.jpg *Source*: http://en.wikipedia.org/w/index.php?title=File:Majinnah4.jpg *License*: unknown *Contributors*: Ministry of Information & Broadcasting, Government of Pakistan.

Image:Jinnah Gandhi.jpg *Source*: http://en.wikipedia.org/w/index.php?title=File:Jinnah_Gandhi.jpg *License*: unknown *Contributors*: Glaurung, Roland zh, Yann

Image:Bagha jatin.JPG *Source*: http://en.wikipedia.org/w/index.php?title=File:Bagha_jatin.JPG *License*: unknown *Contributors*: Dpkpm007, Ragib, Simonxag

File:Ram Prasad Bismil2615.jpg *Source*: http://en.wikipedia.org/w/index.php?title=File:Ram_Prasad_Bismil2615.jpg *License*: unknown *Contributors*: 'Krant 'M.L.Verma (Original photo is still in his possession/library)

Image:Subhas Bose.jpg *Source*: http://en.wikipedia.org/w/index.php?title=File:Subhas_Bose.jpg *License*: unknown *Contributors*: Original uploader was Sidha at en.wikipedia

Image:Hindustan Construction Company logo.svg *Source*: http://en.wikipedia.org/w/index.php?title=File:Hindustan_Construction_Company_logo.svg *License*: unknown *Contributors*: User:Jovianeye

File:Increase2.svg *Source*: http://en.wikipedia.org/w/index.php?title=File:Increase2.svg *License*: unknown *Contributors*: Sarang

Image:Indian Rupee symbol.svg *Source*: http://en.wikipedia.org/w/index.php?title=File:Indian_Rupee_symbol.svg *License*: unknown *Contributors*: User:Orionist

Image: Western-Ghats-Matheran.jpg *Source*: http://en.wikipedia.org/w/index.php?title=File:Western-Ghats-Matheran.jpg *License*: unknown *Contributors*: User:Nichalp

Image: India topo big.jpg *Source*: http://en.wikipedia.org/w/index.php?title=File:India_topo_big.jpg *License*: unknown *Contributors*: Jodelet2, LX, PawełS, Slomox, Sting, 2 anonymous edits

Image:Anamudi 1.jpg *Source*: http://en.wikipedia.org/w/index.php?title=File:Anamudi_1.jpg *License*: unknown *Contributors*: User:Arunguy2002

File:Shola Grasslands and forests in the Kudremukh National Park, Western Ghats, Karnataka.jpg *Source*: http://en.wikipedia.org/w/index.php?title=File:Shola_Grasslands_and_forests_in_the_Kudremukh_National_Park,_Western_Ghats,_Karnataka.jpg *License*: unknown *Contributors*: Karunakar Rayker

File:western ghats section.jpg *Source*: http://en.wikipedia.org/w/index.php?title=File:Western_ghats_section.jpg *License*: unknown *Contributors*: jrsanthosh

Image:Mighty Jog.jpg *Source*: http://en.wikipedia.org/w/index.php?title=File:Mighty_Jog.jpg *License*: unknown *Contributors*: Kiran Sagara(http://new.photos.yahoo.com/kiranfromsagar/albums)

File:Westghats India Monsoon.jpg *Source*: http://en.wikipedia.org/w/index.php?title=File:Westghats_India_Monsoon.jpg *License*: unknown *Contributors*: User:Hinterkappelen

Image:Western-Ghats-rainfall.svg *Source*: http://en.wikipedia.org/w/index.php?title=File:Western-Ghats-rainfall.svg *License*: unknown *Contributors*: User:Nichalp

Image:Nilgiri-Hills.jpg *Source*: http://en.wikipedia.org/w/index.php?title=File:Nilgiri-Hills.jpg *License*: unknown *Contributors*: FlickreviewR, Nichalp, Roland zh, Vadakkan, 1 anonymous edits

Image:Western Ghats RYM.jpg *Source*: http://en.wikipedia.org/w/index.php?title=File:Western_Ghats_RYM.jpg *License*: unknown *Contributors*: Pragadeeshraja

File:Mangalore 038.jpg *Source*: http://en.wikipedia.org/w/index.php?title=File:Mangalore_038.jpg *License*: unknown *Contributors*: Aktron, 1 anonymous edits

Image:Thiruvananthapuram Ponmudi.jpg *Source*: http://en.wikipedia.org/w/index.php?title=File:Thiruvananthapuram_Ponmudi.jpg *License*: unknown *Contributors*: User:Binoyjsdk

Image:Western Ghats.JPG *Source*: http://en.wikipedia.org/w/index.php?title=File:Western_Ghats.JPG *License*: unknown *Contributors*: Tatiraju.rishabh

File:Munnar_tea_gardens.jpg *Source*: http://en.wikipedia.org/w/index.php?title=File:Munnar_tea_gardens.jpg *License*: unknown *Contributors*: Omnipotent, Roland zh, Sreejithk2000, Wouterhagens, 2 anonymous edits

File:Magnify-clip.png *Source*: http://en.wikipedia.org/w/index.php?title=File:Magnify-clip.png *License*: unknown *Contributors*: User:Erasoft24

Image:Keeriparai - View of Hills.JPG *Source*: http://en.wikipedia.org/w/index.php?title=File:Keeriparai_-_View_of_Hills.JPG *License*: unknown *Contributors*: Original uploader was Infocaster at en.wikipedia

File:Lightmatter lion-tailed macaque.jpg *Source*: http://en.wikipedia.org/w/index.php?title=File:Lightmatter_lion-tailed_macaque.jpg *License*: unknown *Contributors*: Fabien1309, Guety, Kersti Nebelsiek, LeaMaimone, Nehrams2020, Petwoe, Quadell, UtherSRG, 1 anonymous edits

File:IndianTiger DK BhadraWLS.jpg *Source*: http://en.wikipedia.org/w/index.php?title=File:IndianTiger_DK_BhadraWLS.jpg *License*: unknown *Contributors*: Dineshkannambadi

File:Doppelhornvogel-09.jpg *Source*: http://en.wikipedia.org/w/index.php?title=File:Doppelhornvogel-09.jpg *License*: unknown *Contributors*: Attis1979, Factumquintus, Hydrargyrum, Kilom691, Rüdiger Wölk

File:Columba elphinstonii.jpg *Source*: http://en.wikipedia.org/w/index.php?title=File:Columba_elphinstonii.jpg *License*: unknown *Contributors*: Atanu Mondal

File:Great-Hornbill.jpg *Source*: http://en.wikipedia.org/w/index.php?title=File:Great-Hornbill.jpg *License*: unknown *Contributors*: User:Kalyanvarma

File:SnailWynaad.jpg *Source*: http://en.wikipedia.org/w/index.php?title=File:SnailWynaad.jpg *License*: unknown *Contributors*: User:Shyamal

File:Malabar Barbet.jpg *Source*: http://en.wikipedia.org/w/index.php?title=File:Malabar_Barbet.jpg *License*: unknown *Contributors*: Sandeep

Image:Sleepingbeautycastle50.jpg *Source*: http://en.wikipedia.org/w/index.php?title=File:Sleepingbeautycastle50.jpg *License*: unknown *Contributors*: Lyght

Image:Canadas_Wonderland_Wonder_Mountain.jpg *Source*: http://en.wikipedia.org/w/index.php?title=File:Canadas_Wonderland_Wonder_Mountain.jpg *License*: unknown *Contributors*: Waldenbg

image:Cinderella Castle Magic Kingdom 07 Edited.jpg *Source*: http://en.wikipedia.org/w/index.php?title=File:Cinderella_Castle_Magic_Kingdom_07_Edited.jpg *License*: unknown *Contributors*: User:Jazzmanluc5

Image:Europapark Rust.JL.jpg *Source*: http://en.wikipedia.org/w/index.php?title=File:Europapark_Rust.JL.jpg *License*: unknown *Contributors*: Florent Pécassou, Jonaslange, Schlurcher

Image:Aerial Canadas Wonderland May 2011.jpg *Source*: http://en.wikipedia.org/w/index.php?title=File:Aerial_Canadas_Wonderland_May_2011.jpg *License*: unknown *Contributors*: airwaves1

Image:Hopi Hari(1).jpg *Source*: http://en.wikipedia.org/w/index.php?title=File:Hopi_Hari(1).jpg *License*: unknown *Contributors*: User:Charlesblack

Image:MGM Dizzee World.jpg *Source*: http://en.wikipedia.org/w/index.php?title=File:MGM_Dizzee_World.jpg *License*: unknown *Contributors*: Sankar Pandian

Image:Vauxhall Gardens by Samuel Wale c1751.jpg *Source*: http://en.wikipedia.org/w/index.php?title=File:Vauxhall_Gardens_by_Samuel_Wale_c1751.jpg *License*: unknown *Contributors*: Samuel Wale

Image:Ferris-wheel.jpg *Source*: http://en.wikipedia.org/w/index.php?title=File:Ferris-wheel.jpg *License*: unknown *Contributors*: Not given

File:Steel Pier 1910s edit.jpg *Source*: http://en.wikipedia.org/w/index.php?title=File:Steel_Pier_1910s_edit.jpg *License*: unknown *Contributors*: Chick Bowen

File:Dorney-park-night-1950.jpg *Source*: http://en.wikipedia.org/w/index.php?title=File:Dorney-park-night-1950.jpg *License*: unknown *Contributors*: User:Bwmoll3

Image:Derelict amusement park in Pemba, Tanzania.JPG *Source*: http://en.wikipedia.org/w/index.php?title=File:Derelict_amusement_park_in_Pemba,_Tanzania.JPG *License*: unknown *Contributors*: User:Loranchet

File:Calico train.JPG *Source*: http://en.wikipedia.org/w/index.php?title=File:Calico_train.JPG *License*: unknown *Contributors*: User:Bobjgalindo

Image:Dorney Park Steel Force Thunderhawk.jpg *Source*: http://en.wikipedia.org/w/index.php?title=File:Dorney_Park_Steel_Force_Thunderhawk.jpg *License*: unknown *Contributors*: Arsonwinter

File:Melbourne Luna Park at Dusk.jpg *Source*: http://en.wikipedia.org/w/index.php?title=File:Melbourne_Luna_Park_at_Dusk.jpg *License*: unknown *Contributors*: User:Donaldytong

Image:Terminator and Paratrooper, night.jpg *Source*: http://en.wikipedia.org/w/index.php?title=File:Terminator_and_Paratrooper,_night.jpg *License*: unknown *Contributors*: Freddo, JackyR, Man vyi, Solipsist, Takeaway

Image:Behemoth (Canada's Wonderland) 06.jpg *Source*: http://en.wikipedia.org/w/index.php?title=File:Behemoth_(Canada's_Wonderland)_06.jpg *License*: unknown *Contributors*: http://www.flickr.com/photos/rollercoasterusa59/

Image:Roller coaster picture dibujo montaña rusa.png *Source*: http://en.wikipedia.org/w/index.php?title=File:Roller_coaster_picture_dibujo_montaña_rusa.png *License*: unknown *Contributors*: Alton, Freddo, Helt, Magnus Colossus, YolanC, 2 anonymous edits

Image:VBRITTO-amusement-stand-louvre.jpg *Source*: http://en.wikipedia.org/w/index.php?title=File:VBRITTO-amusement-stand-louvre.jpg *License*: unknown *Contributors*: User:Vbritto

Image:ISU Central Campus ASLA plaque.jpg *Source*: http://en.wikipedia.org/w/index.php?title=File:ISU_Central_Campus_ASLA_plaque.jpg *License*: unknown *Contributors*: by Cburnett

MIX
Papier aus verantwortungsvollen Quellen
Paper from responsible sources
FSC® C105338

Printed by Books on Demand GmbH, Norderstedt / Germany